Agricultural Research
A Growing Global Divide?

Philip G. Pardey
Nienke Beintema
Steven Dehmer
Stanley Wood

Agricultural Science and Technology Indicators Initiative

International Food Policy Research Institute

Washington, D.C.
August 2006

ISBN 0-89629-529-X

Contents

Acknowledgments

This report was developed with funding from the Agricultural Science and Technology Indicators (ASTI) initiative and additional support from the CGIAR Science Council and the Department of Applied Economics and the International Science and Technology Practice and Policy (InSTePP) center at the University of Minnesota. The authors thank Liliane Ndong and Kate Sebastian for their excellent help in preparing this publication and Howard Elliott for reviewing and commenting on a draft version.

Preface

Sustained, well-targeted, and effectively used investments in R&D have reaped handsome rewards from improved agricultural productivity and cheaper, higher quality foods and fibers. As we begin a new millennium, the global patterns of investments in agricultural R&D are changing in ways that may have profound consequences for the structure of agriculture worldwide and the ability of poor people in poor counties to feed themselves.

This report documents and discusses these changing investment patterns, highlighting developments in the public and private sectors. It revises and carries forward to 2000 data that were previously reported in the 2001 IFPRI Food Policy Report *Slow Magic: Agricultural R&D a Century After Mendel.* Some past trends are continuing or have come into sharper focus, while others are moving in new directions not apparent in the previous series. In addition, this report illustrates the use of spatial data to analyze spillover prospects among countries or agroecologies and the targeting of R&D to address specific production problems like drought-induced production risks. More detailed data on the agricultural research investment trends summarized here can be accessed at www.asti.cgiar.org.

Total Science Spending

Throughout the 20th century, improvements in agricultural productivity have considerably alleviated poverty and starvation and fueled economic progress. Further, a large body of evidence closely links productivity improvements to investments in agricultural research and development (R&D).[1] In the past several decades, however, many countries have made major changes in the way they fund and organize public agricultural R&D and the incentives affecting private R&D. These changes are reflected in the shifting patterns of support for agricultural R&D, reported here, raising questions about the prospects for sustaining productivity growth over the next several decades and beyond.

Agricultural R&D is not conducted in isolation from the rest of science.[2] Agricultural scientists have a long history of drawing on and adapting findings from the basic biological, chemical, and other sciences to further their own research, and scientific spillovers have flowed in the other direction as well. Moreover, given contemporary developments, particularly in the genetic and informational sciences, the boundaries between agriculture and other sciences are increasingly becoming blurred. Consequently, putting the agricultural sciences in the context of overall science spending is instructive.

In 2000, $731 billion was invested in *all* the sciences worldwide,[3] including research conducted by both public agencies and private firms. This represented about 1.7 percent of the world's $42.4 trillion gross domestic product (GDP) that year, and an increase of nearly one-third over the inflation-adjusted total of just five years earlier (Table 1). Real spending in all regions of the world

Table 1—Total gross domestic expenditures on research and development, 1995 and 2000

	Total R&D expenditures (million 2000 international dollars)		Share of global total (percent)	
Region/country	**1995**	**2000**	**1995**	**2000**
Developing countries				
Asia-Pacific (26)	52,416	94,950	9.3	13.0
China	19,469	48,247	3.5	6.6
India	11,678	20,749	2.1	2.8
Latin America and the Caribbean (32)	17,222	21,244	3.1	2.9
Brazil	9,771	12,398	1.7	1.7
Sub-Saharan Africa (44)	3,008	3,992	0.5	0.5
Middle East and North Africa (18)	8,626	14,893	1.5	2.0
Other developing countries (21)	19,002	21,895	3.4	3.0
Developing-country subtotal (141)	100,274	156,975	17.9	21.5
High-income countries				
Japan	89,964	99,500	16.0	13.6
United States	196,358	263,043	35.0	36.0
High-income country subtotal (23)	461,367	573,964	82.1	78.5
Total (164)	561,641	730,939	100.0	100.0

SOURCES: Based on Pardey, Dehmer, and El Feki (2006) using data from numerous sources.

NOTES: The number of countries included in the regional totals is shown in parentheses. "Other developing countries" includes many Eastern European, former Soviet countries; "Latin America and the Caribbean" includes Mexico, a member of the Organisation for Economic Co-Operation and Development (OECD); "high-income countries" only includes the high-income members of the OECD—thus excluding a number of high-income countries, such as South Korea and French Polynesia (grouped under Asia-Pacific), Bahrain, Israel, Kuwait, Qatar, and United Arab Emirates (grouped under Middle East and North Africa), and the Bahamas (grouped under Latin America and the Caribbean). All data were first compiled in current local currency units, then deflated to 2000 constant currency units, and finally converted to international dollars using purchasing power parity (PPP) exchange rates.

increased between 1995 and 2000, but growth was uneven.[4] Of the developing countries, the most notable increases were in the Asia–Pacific and Middle East and North Africa regions, with hefty increases of 11.9 and 11.5 percent, respectively (the latter fueled by rapid spending increases in Israel and Turkey). While the overall average rate of growth for developing countries was 8.6 percent per year over the 1995–2000 timeframe, regional averages for developing countries ranged from lows of 1.9 percent per year for the "other developing countries" category (which includes several former Soviet states) and 3.0 percent per year for Sub-Saharan Africa, to notable highs of 19.7 percent per year for China and 12.2 percent per year for India.

These regional trends hide a profoundly disturbing reality—evidence of a large and, in places, growing divide between the scientific haves and have-nots. For example, the overall growth in the Asia–Pacific region masks the fact that just two countries, China and India, accounted for 89 percent of the $42.5 billion increase in regional spending from 1995 to 2000. Put another way, China and India accounted for 59 percent of the region's scientific spending in 1995, jumping to 73 percent of the regional total by 2000. In contrast, while research spending in the seven Pacific countries (including Fiji, French Polynesia, New Caledonia, and others) grew by as much as 9.4 percent annually from 1995, this was from an exceptionally small base, so their $120.7 million total in 2000 represents just a minuscule 0.13 percent of the Asia–Pacific region's total science spending.

Although geographically large and home to over 10 percent of the world's population, Sub-Saharan Africa accounts for just 0.5 percent of the world's gross investment in science. Further, South Africa, with less than 7 percent of this region's population, accounts for about two-thirds of the regional total for gross domestic expenditures on R&D. While 39 of the 44 countries in Sub-Saharan Africa for which data are available increased their investments in R&D between 1995 and 2000, South Africa accounted for about 61 percent of the nearly $1 billion increase.

Middle East and North Africa fared a bit better than Sub-Saharan Africa, with a real increase of R&D investment of almost 73 percent between 1995 and 2000. Indeed, the only country tracked in this region that reported a decrease in investment was Kuwait, with a period decline of almost 34 percent. As in Sub-Saharan Africa, however, the growth is highly concentrated, with Israel and Turkey alone accounting for almost 79 percent of the region's increase during this period.

The bifurcation in science spending is widespread, and these new data make the significant geopolitical concentration of science spending worldwide manifestly clear. In 2000, the top five countries (in descending order, the United States, Japan, Germany, France, and the United Kingdom) accounted for 68.6 percent of the world's total science spending, and the two top spending countries alone (the United States and Japan) accounted for 63 percent of the total for Organisation for Economic Co-Operation and Development (OECD) countries.[5] Expanding this group to the top 10 countries—which includes Italy, Canada, the lower income but fast-growing countries China and India, and South Korea—the share comes in at 81.6 percent of the world total. Moreover, the share of the bottom 80 countries (accounting for 11.1 percent of the world's population in 2000 but only 2.4 percent of global GDP) slipped from 0.29 percent of the global total in 1995 to 0.26 percent in 2000. Put together, this is evidence of a large and sustained, if not growing, gap between a comparatively small group of scientific haves and a substantial group of scientific have-nots.

Box 1

INTERNATIONALLY COMPARABLE MEASURES OF R&D

Cross-country comparisons of R&D expenditures, like most international comparisons of economic activity more generally, are confounded by substantial differences in price levels among countries. This is particularly a problem when valuing something like expenditures on agricultural R&D, where typically two-thirds of the expenditures are on local scientists and support staff, not capital or other goods and services that are commonly traded internationally. For example, the average salary received by full professors working at large public universities in the United States (net of benefits) was $88,457 in 2004/05. A comparable annual salary paid to a chief scientific officer in Bangladesh working for the national government's main agricultural research agency was TK 20,700 (equivalent to 1,683 international dollars when converted using purchasing power parities [PPPs] or only US$316 when converted using official exchange rates), while a mid-career senior scientist working for Embrapa, Brazil, earned an average of 72,348 reals (65,705 international dollars or US$30,020).

Converting research expenditures from different countries to a single currency using official exchange rates tends to understate the quantity of research resources used in economies with relatively low prices, while overstating the quantity of resources used in countries with high prices.[a] At present, there is no entirely satisfactory method for comparing consumption or expenditures among countries at different points in time (or for that matter, at the same point in time). Unfortunately, the choice of deflator and currency converter can have substantial consequences for both the measure obtained and its interpretation.

Most of the research expenditures in this report are denominated in 2000 "international dollars" using PPPs to do the currency conversions.[b] For convenience of interpretation, the reference currency—here an international dollar—is set equal to a U.S. dollar in the benchmark year.

Figure B1 contrasts the regional expenditure shares both for public agricultural research expenditures using PPPs versus official exchange rates to do the currency conversion. The left-hand side of the figure denotes 2000 research spending in international dollars obtained using PPPs, while the right-hand side of the figure reports the U.S. dollar estimates obtained using the same underlying R&D data together with official exchange rates. Taking the PPP estimates to be more representative of the amount of resources committed to research, the U.S. dollar estimates overstate the share of developed-country agricultural research in the global total and grossly understate the African, Chinese, and other Asia–Pacific shares.

Figure B1 Agricultural research spending in U.S. versus international dollars, 2000

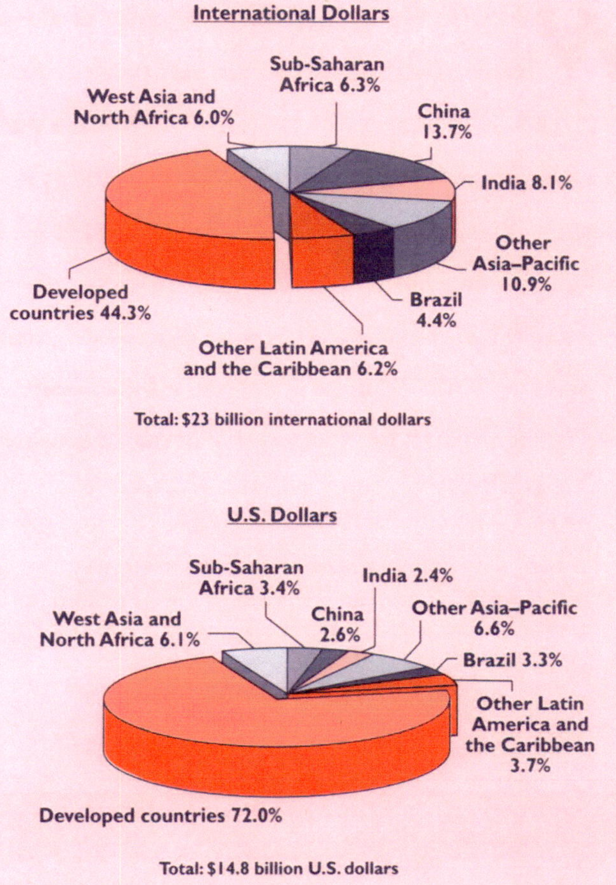

SOURCE: Calculated by authors based on data reported in Table 2.

SOURCES: Pardey, Roseboom, Craig 1992; World Bank 2005b.

[a] *A country's international price level is the ratio of its PPP rate to its official currency exchange rate for U.S. dollars. In other words, the international price level is an index of the costs of goods in one country at the current rate of exchange relative to the costs of the same bundle of goods in a numeraire country, in this case the United States. For example, in 2000 the ratio of PPP to exchange rate for Australia was 0.77, indicating that average prices in Australia were 23 percent lower than they were in the United States. The corresponding ratio for Bangladesh was 0.22, meaning that a bundle of goods and services purchased for $100 in the United States cost only $22 dollars in Bangladesh.*

[b] *We use a procedure described by Pardey, Roseboom, and Craig (1992) that first deflates research expenditures expressed in current local currency units to a base year set of prices (2000, in this case) using a local price deflator and then converts to a common currency unit (specifically, international dollars) using PPPs for 2000 obtained from the World Bank (2005b) rather than the more familiar official exchange rates.*

Public Agricultural R&D

Research Spending Trends

Worldwide, public investments in agricultural research increased by 51 percent in inflation-adjusted terms over the past two decades, from an estimated $15.2 billion (2000 international dollars) in 1981 to $23.0 billion in 2000 (Table 2). These data reveal a significant structural shift: during the 1990s, developing countries as a group undertook more of the world's public agricultural research than the developed countries.[6] The Asia–Pacific region has continued to gain ground, accounting for an ever-larger share of the developing-country total since 1981. Just two countries from this region, China and India, accounted for 39.1 percent of the developing world's expenditure on agricultural R&D in 2000, a substantial increase from their 22.9 percent combined share in 1981. In stark contrast, Sub-Saharan Africa has continued to lose market share, falling from 17.3 to 11.4 percent of the developing-world total between 1981 and 2000.

Paralleling spending patterns for all the sciences, agricultural R&D has become increasingly concentrated in a handful of countries worldwide. Just four countries—the United States, Japan, France, and Germany—accounted for two-thirds of the public research done by rich countries in 2000, about the same as two decades before. Similarly, just five developing countries—China, India, Brazil, Thailand, and South Africa—undertook 53.3 percent of the developing world's public agricultural research in 2000, up from 40

Table 2—Total public agricultural research expenditures by region, 1981, 1991, and 2000

	Agricultural R&D spending (million 2000 international dollars)			Share of global total (percent)		
Region/country	**1981**	**1991**	**2000**	**1981**	**1991**	**2000**
Developing countries						
Asia–Pacific (28)	3,047	4,847	7,523	20.0	24.2	32.7
China	1,049	1,733	3,150	6.9	8.7	13.7
India	533	1,004	1,858	3.5	5.0	8.1
Latin America and the Caribbean (27)	1,897	2,107	2,454	12.5	10.5	10.7
Brazil	690	1,000	1,020	4.5	5.0	4.4
Sub-Saharan Africa (44)	1,196	1,365	1,461	7.9	6.8	6.3
Middle East and North Africa (18)	764	1,139	1,382	5.0	5.7	6.0
Developing-country subtotal (117)	6,904	9,459	12,819	45.4	47.3	55.7
High-income countries						
Japan	1,832	2,182	1,658	12.1	10.9	7.2
United States	2,533	3,216	3,828	16.7	16.1	16.6
High-income country subtotal (22)	8,293	10,534	10,191	54.6	52.7	44.3
Total (139)	15,197	19,992	23,010	100.0	100.0	100.0

SOURCES: Calculated by authors based on Agricultural Science and Technology Indicators (ASTI) initiative data; Pardey and Beintema (2001); RICYT (2005); Casas, Solh, and Hafez (1999); OECD (2005); Eurostat (2005); and USDA/CRIS (2006).

NOTES: The number of countries included in the regional totals is shown in parentheses. See notes to Table 1 regarding country aggregation/groupings. These estimates exclude Eastern Europe and former Soviet Union countries. Regional totals were scaled up from national spending estimates for countries that represented 79 percent of the reported Sub-Saharan African total, 89 percent of the Asia–Pacific total, 86 percent of the Latin America and Caribbean total, 57 percent of the Middle East and North Africa total, and 84 percent of the high-income country total.

Table 3—Spatial concentration of public expenditures in agricultural R&D worldwide, 1995 and 2000

Country grouping	1995 (percent)	2000 (percent)	2000–02 (percent) GDP	Population	Agricultural land	Agricultural production Crops	Livestock	Total
Top 5	47.5	50.0	52.6	51.8	22.7	38.6	42.8	40.4
Top 10	61.7	62.4	66.5	56.1	33.2	52.8	54.2	53.4
Bottom 80	8.6	6.3	5.7	11.3	13.6	7.1	3.9	5.8

SOURCES: Calculated by authors based on Agricultural Science and Technology Indicators (ASTI) initiative data and World Bank (2006).

NOTES: The top 10 agricultural R&D expenditure countries in 1995 (in descending order) were United States, Japan, China, India, Brazil, Germany, South Korea, Australia, United Kingdom, and France; the top 10 countries in 2000 (in descending order) were United States, China, India, Japan, Brazil, Germany, Australia, South Korea, United Kingdom, and Canada. GDP and population data are from 2000; agricultural production and land area data are from 2002.

percent in 1981.[7] Meanwhile, only 6.3 percent of agricultural R&D worldwide was conducted in 80 (mainly low-income) countries—home to some 625 million people in 2000 and accounting for nearly 14 percent of the world's agricultural land area. Notably, this 80-country share of global agricultural R&D spending is slightly more than their corresponding value share (5.8 percent) of worldwide agricultural output (Table 3).

A shifting and widely disbursed pattern of growth is evident among regions (Figure 1). Certainly, the more recent rates of increase in inflation-adjusted spending for all developing regions of the world failed to match the rapid ramping up of public agricultural R&D spending of the 1970s (Pardey and Beintema 2001). The growth in spending for the Asia–Pacific region held strong, averaging 4.3 percent per year in the 1980s and 3.9 percent per year in the decade to follow. Growth in China and India picked up in the late 1990s, in both instances reflecting government policies to revitalize public research and improve its commercialization prospects—including linkages with the private sector.[8] Spending growth throughout the Latin American region as whole was more robust during the 1990s than the 1980s, although the recovery was more fragile and less certain for some countries in the region (such as Brazil, where rates of spending contracted at the close of the 1990s, then partially recovered in 2000/01).

Overall investments in agricultural R&D in Sub-Saharan Africa failed to grow by more than 1 percent per year during the 1990s—the continuation of a longer run slowdown. Even more disturbing, about half of the 27 African countries for which national estimates were available spent less on agricultural R&D in 2000 than they did in 1991 (Beintema and Stads 2004).

A notable feature of the growth trends is the contraction in support for public agricultural R&D among rich countries (Figure 1). During the 1980s, real public agricultural R&D spending grew by an average of 2.3 percent per year for the rich countries compared with an average rate of decline of 0.6 percent per year during the 1990s. While spending in the United States picked up in the second half of the 1990s (2.9 percent per year for 1995–2000 versus 1.5 percent per year for 1990–95), a massive reduction in public research funding

Figure 1 Public agricultural R&D spending trends

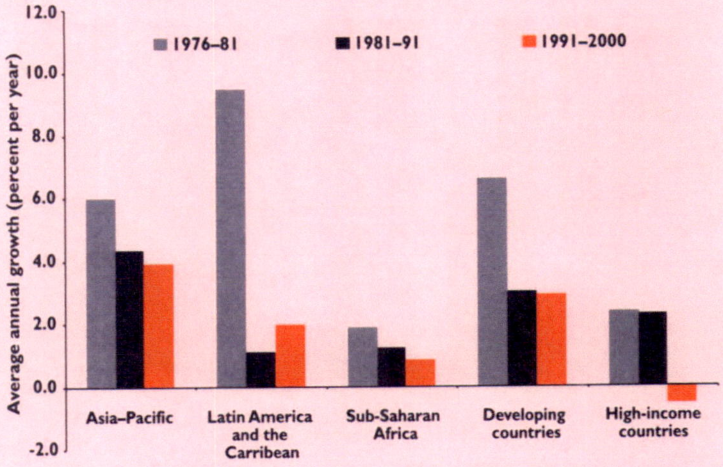

SOURCE: Table 2.

NOTE: Inflation-adjusted growth rates were calculated as weighted regional averages, using the least-squares method described in World Bank (2006, 305).

occurred in Japan (and, to a lesser degree, several European countries) toward the end of the 1990s, leading to a decline (albeit small) in rich-country spending as a whole for the decade. Once again, these new data reinforce the longer run trends observed earlier—namely, a fairly widespread scaling back, or at best a slowing down of support for publicly performed agricultural research among rich countries. In part, this points to a shifting emphasis from publicly to privately performed agricultural R&D, and to a shift in government spending priorities.

Inevitably, this will affect productivity prospects in agriculture for the countries in question. In addition, as Pardey, Alston, and Piggott (2006) suggest (and as is discussed in more detail later in this report), a more subtle and arguably more important consequence is that slowdowns or cutbacks in rich-country spending will curtail the future spillovers of ideas and new technologies from rich to poor countries. These rich–poor country linkages will be even more attenuated as the funding trends proceed in parallel with other policy and market developments, like strengthening intellectual property rights and biosafety regulations and a reorientation of rich-country R&D away from productivity gains in food staples toward concerns over the environmental effects of agriculture, as well as the food quality, medical, energy, and industrial applications of agricultural commodities. While this research is likely to generate substantial economic value, the fact that developed countries, as a group, still account for nearly 41 percent of public agricultural R&D worldwide (and almost 80 percent of all science spending) means the consequences of such continued funding, policy, and market trends could be particularly pronounced in terms of the productivity-enhancing effects on food staples.

The broad trends documented here mask many of the aspects of agricultural R&D funding that have important practical consequences. For example, undue variability in research funding continues to be problematic for many developing-country research agencies. This is especially troubling for agricultural R&D, given the long gestation period for new crop varieties and livestock breeds and the desirability of long-term employment assurances for scientists and other staff (Pardey, Alston, and Piggott 2006). Variability encourages an overemphasis on short-term projects or those with short lags between investment and outcomes, and adoption. It also discourages specialization of scientists and other resources in areas of work where sustained funding may be uncertain, even when these areas have high payoff potentials.

Institutional Orientation

In this report, public agricultural research includes research performed by government, higher education, and nonprofit agencies.[9] There are substantial differences among countries and between regions in the structure of the public research sector (Figure 2). Public research in the United States is done mainly in state agricultural experiment stations (SAES) located principally in colleges of agriculture and in federally administered, but often regionally located, labs of the United States Department of Agriculture (USDA). The SAES share of total USDA-SAES research has increased over the past several decades, from 67.2 percent in 1980 to 73.6 percent in 2004. Notably, state government financing of SAES-performed R&D has slipped from 54 percent in 1980 to 40 percent in 2004. Federal funding of SAES-performed research has picked up in more recent years—including funds disbursed through USDA, as well as those from a host of other federal government agencies (like the National Institutes of Health, Department of Defense, and Environmental Protection Agency)—signaling a substantial diversification of what now constitutes "agricultural" research. Funds from other (often private or self-generated) sources have also increased, including royalty revenues and licensing income from protected intellectual property.

A much larger share of public agricultural R&D in Latin America is conducted by government agencies—about 74 percent of

Figure 2 Institutional orientation of public agricultural R&D, 1981–2000

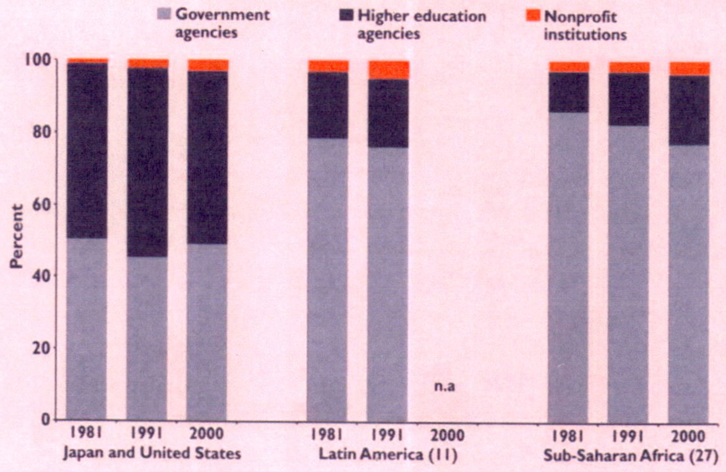

SOURCE: Calculated by authors based on Agricultural Science and Technology Indicators (ASTI) initiative data.

NOTES: The number of countries included in each category is shown in parentheses. The reported shares for Japan and the United States may understate the role of nonprofit institutions. n.a. indicates not available.

the total in 1996 (the latest year for which an 11-country total is available). This is similar to the government agency share in our 27-country Sub-Saharan African total. Like Latin America, a small but growing proportion of public research in Sub-Saharan Africa is conducted by nonprofit institutions; in 2000, for example, they accounted for 3 percent of total agricultural research staff (Figure 2). Nonprofit institutions are often managed by independent boards not directly under government control. Many are closely linked to producer organizations from which they receive the lion's share of their funding, typically by way of taxes levied on production or exports. Examples include agencies conducting research on tea (Kenya, Tanzania, Malawi), coffee (Uganda, Kenya, Tanzania), cotton (Zambia), and sugar (Mauritius, South Africa). Noteworthy is the establishment of various other forms of nonprofit institutions, not linked to producer organizations, in a number of countries, such as Madagascar and Togo.

In 2000, of the full-time equivalent (fte) researchers working in nonprofit institutions in Sub-Saharan Africa, Madagascar, Mauritius, and South Africa (the southern African region) employed about three-quarters. Togo was the only country in West Africa to employ researchers in nonprofit agencies, according to the available data, but they totaled only 9 fte researchers in 2001. Although the number of fte researchers working for nonprofit agencies throughout Sub-Saharan Africa has increased considerably, the rate of growth was less than in the corresponding government and higher education sectors, with the result that the nonprofit researcher share was smaller in 2000 than three decades earlier.

The continuing scarcity and comparatively small size of national agricultural science institutions throughout Africa has spurred attempts to strengthen subregional research coordination and implementation capacities in western, eastern, and southern Africa.[10] These regional efforts were conceived to stimulate knowledge and technology spillovers among countries within regions, improve the capacity to search for and obtain access to new knowledge and technologies from further afield, achieve economies of scope and scale in the conduct of research, coalesce a critical mass of local scientific expertise around regional priorities (which are not necessarily the sum of national priorities), and achieve these aims in the face of persistent national funding vagaries and the ravages of HIV/AIDS on scientific capacity within the region.

Typically, research activities have been organized as (sub)regional research networks coordinated by a regional scientific research organization (for example, the Association for Strengthening Agricultural Research in East and Central Africa [ASARECA]). A good number of these networks were originally established and managed by centers of the Consultative Group on International Agricultural Research (CGIAR or CG). For many of the same reasons behind the subregional networks developed around national research capacities, the CGIAR is also in the process of reconstituting its own African efforts through joint subregional programs between relevant CG centers, scientific research organizations, and national research agencies. Integrated CGIAR medium-term investment plans for eastern and central Africa, for example, are now well advanced.

The extent to which these new institutional arrangements have increased funding for Sub-Saharan African research is unclear, if not questionable.[11] Even in the absence of increased funding, the hope is that these regionalized arrangements will improve the relevance and harmonization of R&D efforts sufficient to realize more cost-effective research (via enhancing spillovers, achieving a critical mass, and reducing R&D lag times), thereby increasing the regional *and* national benefits from research. A pessimistic view is that these regional arrangements serve merely to redirect money otherwise committed to national and international research, while at the same time increasing transaction costs and the earmarking of research funds in ways that undermine research efficiencies. Indeed, it is unclear if many of the arrangements already in place (or those contemplated) substantially alter the existing incentives to innovate (and to mobilize funding for that innovation), such that the same problems that gave rise to an underfunding of national research will simply be compounded by strategic behavior among agencies now also operating in regional institutional frameworks.[12]

Research Intensities

Turning now from absolute to relative measures of R&D investments, developed countries as a group spent $2.36 on public agricultural R&D for every $100 of agricultural output in 2000, a sizable increase over the $1.41 they spent per $100 of output two decades earlier but, notably, slightly down from the 1991 estimate of $2.38 (Figure 3). This longer run rise in research intensity starkly contrasts with the group of developing countries, where since 1981 there has been no measurable growth in the intensity of agricultural research (that is, agricultural R&D spending expressed as a percentage of agricultural gross domestic product [AgGDP]). In 2000, the developing world spent just 53 cents on public agricultural R&D for every $100 of agricultural output.

At first glance, the combined rise in rich-country intensity ratios and the stagnating research intensities for poor countries belies the evidence presented in Figure 1, where the growth in overall investments in agricultural

Figure 3 Intensity of public agricultural R&D

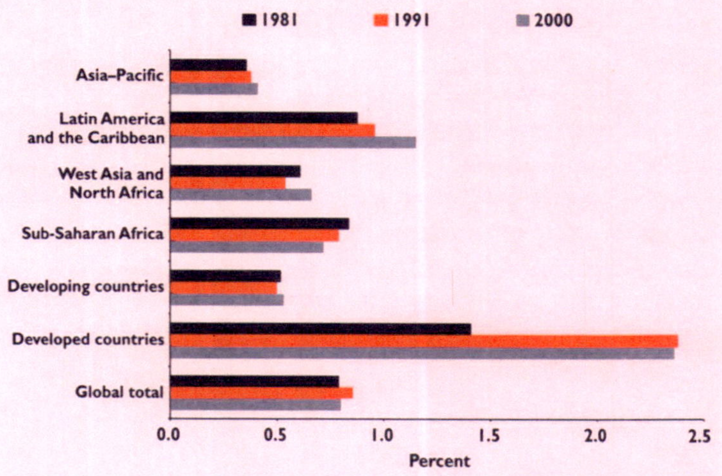

SOURCE: Calculated by authors based on Agricultural Science and Technology Indicators (ASTI) initiative data. Agricultural GDP data are from World Bank (2005b).

NOTES: The intensity ratios measure total public agricultural R&D spending as a percentage of agricultural output agricultural GDP. The developing-country category includes countries that also constitute regional totals.

R&D in poor countries (3.1 percent per year from 1981 to 2000) significantly outpaced the rise in spending by rich countries (1.1 percent per year). Delving deeper, agricultural output grew much faster in aggregate for developing versus developed countries over the past several decades, so that the faster growth in aggregate agricultural research spending among poor countries has, nonetheless, barely kept pace with the corresponding growth in output. In other words, the scientific or knowledge intensity of agricultural production grew at a much faster rate in rich relative to poor countries; indeed, the intensity gap has grown over the past several decades.[13] In addition, more than half of the developed countries for which data were available had higher research intensity ratios in 2000 than they did in 1981, and the majority of them spent in excess of $2.30 on public agricultural R&D for every $100 of AgGDP. However, only 10 of the 26 Sub-Saharan countries in our sample had higher 2000 intensity ratios than in 1981, although most countries in our Asian and Latin American samples (9 of 11 Asian countries and 7 of 11 Latin American countries) increased their intensity ratios over the 1981–2000 period.[14]

Other research intensity ratios are also revealing (Table 4). Rich countries spent $692 per agricultural worker in 2000, more than double the corresponding 1981 ratio. Poor countries spent just $10 per agricultural worker in 2000, substantially less than double the 1981 figure. These rich/poor country differences are, perhaps, not too surprising. A much smaller share of the rich-country workforce is employed in agriculture, and the absolute number of agricultural workers declined more rapidly in rich countries than it did in the poor ones.

While only some segments of society are directly involved in agriculture as producers, everyone consumes agricultural outputs, and so a look at agricultural R&D spending per capita is instructive. These new data signal a break from earlier trends. For rich countries, spending per capita rose substantially from 1981 to 1991 (a continuation of earlier trends documented by Pardey and

Table 4—Alternative public agricultural research intensities, 1981, 1991, and 2000

	Agricultural R&D spending (2000 international dollars)					
	Per capita			Per capita of economically active agricultural population		
Region/grouping	**1981**	**1991**	**2000**	**1981**	**1991**	**2000**
Asia–Pacific	**1.31**	**1.73**	**2.35**	**3.84**	**5.23**	**7.57**
Latin America and the Caribbean	**5.43**	**4.94**	**4.96**	**45.10**	**50.54**	**60.11**
Sub-Saharan Africa	**3.14**	**2.69**	**2.28**	**9.79**	**9.04**	**8.22**
Middle East and North Africa	**3.24**	**3.63**	**3.66**	**19.15**	**27.30**	**30.24**
Developing-country subtotal	**2.09**	**2.34**	**2.72**	**6.91**	**8.14**	**10.19**
High-income country subtotal	**10.91**	**13.04**	**11.92**	**316.52**	**528.30**	**691.63**
Total	**3.75**	**4.12**	**4.13**	**14.83**	**16.92**	**18.08**

SOURCE: Calculated by authors based on Agricultural Science and Technology Indicators (ASTI) initiative data. Population and economically active agricultural population are from FAO (2005a and b).

Beintema 2001), but declined thereafter so that spending per capita in 2000 had slipped well below 1991 levels. This rich-country reversal was driven mainly by developments in Japan; although, only half the developed countries continued to increase their per capita spending on agricultural R&D throughout the 1990s.

Spending per capita levels are much lower among poor countries compared with the rich ones. Developing countries (especially those in Africa) typically spent less than $3 per capita in 2000, whereas 59 percent of the developed countries invested more than $10 per capita. Nonetheless, and in contrast to the group of rich countries, per capita spending for the group of poor countries continued to rise, albeit slowly, from $2.12 in 1981 to $2.72 in 2000. The outlier to this general trend is Sub-Saharan Africa, where agricultural R&D spending per capita has continued to decline since at least 1981.[15]

Figure 4 Spending per researcher, 1981 to 2000

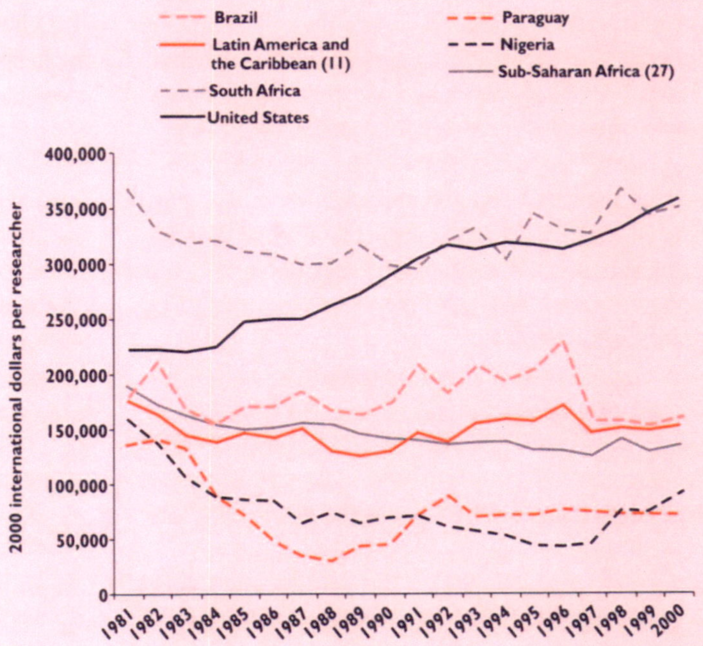

SOURCE: Calculated by authors based on Agricultural Science and Technology Indicators (ASTI) initiative data and, for the United States, USDA/CRIS (2006).

NOTE: The number of countries in the regional categories is shown in parentheses.

Spending per Scientist

Agricultural R&D spending grew comparatively quickly in many parts of the developing world during the 1980s, but not fast enough to outpace the corresponding growth in the number of scientists. After adjusting for inflation and cross-country differences in price levels, average spending per fte scientist for a sample of 27 Sub-Saharan African countries fell quite markedly from $190,000 (2000 international dollars) in 1981 to $132,000 two decades later (Figure 4). The fall was less pronounced, but evident nonetheless, for the 11 Latin American and Caribbean countries for which time-series data were available. Throughout the 1990s, real spending per scientist stabilized somewhat, although it drifted down slightly in Sub-Saharan Africa with signs of a slight recovery in Latin America. These regional averages inevitable obscure important country-specific details. For example, the 1980s decline in support per scientist was much more pronounced in Nigeria and Paraguay compared with their respective regional averages.

Developing-country patterns (with a few exceptions) markedly contrast developments in the United States. U.S. public-sector spending per fte grew steadily from an average of $222,017 (2000 prices) in 1981 to $356,911 in 2000—a 2.6 percent increase per year in the real resources available per researcher. Of course, not all states tracked the U.S. average trend. Spending per scientist in some states (including Delaware, Pennsylvania, and South Carolina) grew by less than 1 percent per year, while 10 states grew faster than 4 percent annually. Moreover, after adjusting for price-level differences between countries, spending per scientist in some developing countries (such as South Africa) are comparable with U.S. levels.

Public Versus Private Agricultural R&D

For almost all of agriculture's 10,000 year history, innovation was mainly a private, individual undertaking. Improved crop varieties, livestock breeds, and farm management practices were typically the result of farmer experimentation—adapting and developing earlier ideas, then passing on inventions to siblings, children, and fellow farmers. Collectively conceived and funded public research did not begin until the early to mid-1700s as part of the efforts of the agrarian societies that formed throughout the United Kingdom and Europe at that time. From these institutional roots, the publicly funded and operated agricultural experiment stations developed around the mid-1800s. But even as public agricultural R&D took off, private agricultural R&D continued to flourish. It too evolved, from the tinkering and trial-and-error efforts of many individuals—most operating alone—to large-scale input supply firms investing in their own private R&D facilities. For example, in U.S. agriculture alone, Eli Whitney patented the cotton gin, Cyrus McCormick's mechanical reaper "made bread cheap," John Deere's steel-tipped moldboard plows helped tame the prairies, and Hiram Moore built the first combined harvester (combining a reaper and a thresher in one machine). The list of biological innovators is less well-known, but the legendary Luther Burbank—who developed scores of new and improved varieties, many of which still bear his name—is representative of thousands of farmer-scientists who, by careful selection and in some cases hybridization, improved the plant varieties available to American farmers.[16]

Particularly in agriculture, however, it is difficult for individuals to fully appropriate the returns from their research investments, and it is widely held that some government action is warranted to ensure an adequate investment in R&D to fully capture the public good (Pardey, Alston and Piggott 2006). The private sector has continued to emphasize inventions that are amenable to various intellectual property protection options such as patents, and more recently, plant breeders' rights and other forms of intellectual property.[17] Private investments in agricultural R&D, like investments in all forms of research, are motivated and sustained by the returns to innovation reaped by that investment. Intellectual property policies and practices are but one dimension of the incentive to innovate. Potential market size and the cost of servicing the market—in turn dependent on the state of communication and transportation infrastructure, farm structure and size, and farm income—are important dimensions as well. So too is the pattern of food consumption. As incomes rise, larger shares of the food expenditures go toward food processing, convenience, and other attributes of food—areas where significant shares of private agricultural research effort are directed.

A large private presence is evident in agricultural R&D, but with dramatic differences between rich and poor, and among individual, countries (Table 5). In 2000, global spending on agricultural R&D (including prefarm-, onfarm-, and postfarm-oriented R&D) was $36.0 billion—about 36 percent of which was performed by private firms and the remaining 64 percent by public agencies. Notably, about 93 percent of that private R&D was performed in rich countries, where some 54 percent of the agricultural R&D is private. In developing countries, only 6 percent of the agricultural R&D is private and there are large disparities in the private share among regions of the developing world. In the Asia–Pacific region, nearly 8 percent of the agricultural R&D is private compared with only 2 percent of the research throughout Sub-Saharan Africa.

The majority of private R&D in Sub-Saharan Africa was oriented to crop-improvement research, often (but not always) dealing with export crops, such as cotton (in Zambia and Madagascar) and sugarcane (in Sudan and Uganda). Virtually all the firms are small, both in terms of total spending and numbers of researchers. They involve a mix of locally owned companies (for example, Pannar Seeds in Greytown, South Africa, or Kenana Sugar Company in Sudan), as well as local affiliates of multinational companies. Moreover, almost two-thirds of the private research performed throughout the whole region was done in South Africa. Given the tenuous market realities facing much of African agriculture, it is unrealistic to expect marked and rapid development of locally conducted private R&D. That said, there is substantial potential, perhaps, for tapping into private agricultural R&D done elsewhere—maybe through creative public–private joint ventures.

Table 5—Estimated global public and private agricultural R&D investments, circa 2000

	Expenditures (million 2000 international dollars)			Share (percent)	
Region/country	**Public**	**Private**	**Total**	**Public**	**Private**
Asia–Pacific	7,523	663	8,186	91.9	8.1
Latin America and the Caribbean	2,454	124	2,578	95.2	4.8
Sub-Saharan Africa	1,461	26	1,486	98.3	1.7
Middle East and North Africa	1,382	50	1,432	96.5	3.5
Developing-country subtotal	12,819	862	13,682	93.7	6.3
High-income country subtotal	10,191	12,086	22,277	45.7	54.3
Total	23,010	12,948	35,958	64.0	36.0

SOURCE: Calculated by authors based on Agricultural Science and Technology Indicators (ASTI) initiative data and data presented in OECD (2005).

Rich-country agricultural R&D is increasingly a private-sector pursuit. The privately performed share of agricultural R&D in OECD countries grew steadily from 43.6 percent in 1981 to 54.3 percent in 2000 (Table 6). This trend may well continue if the science of agriculture increasingly looks like the sciences more generally. In the United States, for example, the private sector conducted nearly 52 percent of agricultural R&D in 2000 compared with 72 percent of all R&D expenditures that same year (NSF 2005).

The rich/poor country disparity in the intensity of agricultural research (noted in Figure 3) is magnified dramatically if private research is also factored in (Figure 5). In 2000, developing countries as a group had an agricultural R&D intensity ratio of 0.56 percent (that is, for every $100 of agricultural GDP, 56 cents was spent on agricultural R&D) compared with a ratio of 5.16 percent for developed countries. This results in a rich- versus poor-country intensity ratio of 9.2:1 compared with a 4.5:1 ratio if just public research spending were considered.

Figure 5 Public, private, and total agricultural R&D intensities, circa 2000

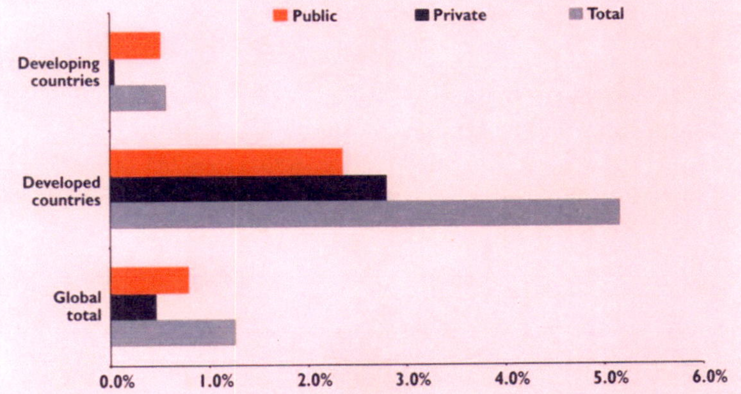

SOURCE: Calculated by authors based on Agricultural Science and Technology Indicators (ASTI) initiative data and data presented in OECD (2005).

NOTE: The intensity ratios measure total public and private agricultural R&D spending as a percentage of agricultural output (agricultural GDP).

Table 6—Total public and private agricultural R&D expenditures, selected countries from the Organization for Economic Co-Operation and Development, 1981–2000

Country	Private share of total (percent)			Average annual growth rate (percent per year)	
	1981	1991	2000	1981–91	1991–2000
Australia	5.9	22.0	24.8	15.3	4.0
Canada	17.3	21.5	34.0	2.5	5.5
France	44.1	52.0	74.7	8.2	2.7
Germany	56.2	43.6	53.6	2.4	0.7
Japan	36.6	48.4	58.6	7.5	1.8
The Netherlands	44.8	56.1	57.7	9.3	1.1
United Kingdom	55.9	66.8	71.5	6.0	1.7
United States	49.3	51.0	51.5	3.6	2.4
OECD total (22)	43.6	48.5	54.3	5.2	2.1

SOURCE: Calculated by authors based on data presented in OECD (2005).

NOTES: Average annual growth rates calculated using the least-squares regression method, as described by the World Bank (2006, 305). In 1981, private sector agricultural R&D spending was estimated to be $6,422 million (2000 international dollars), $9,930 million in 1991, and $12,086 million in 2000.

Figure 6 Distribution of the world's cultivated systems by agroecological class

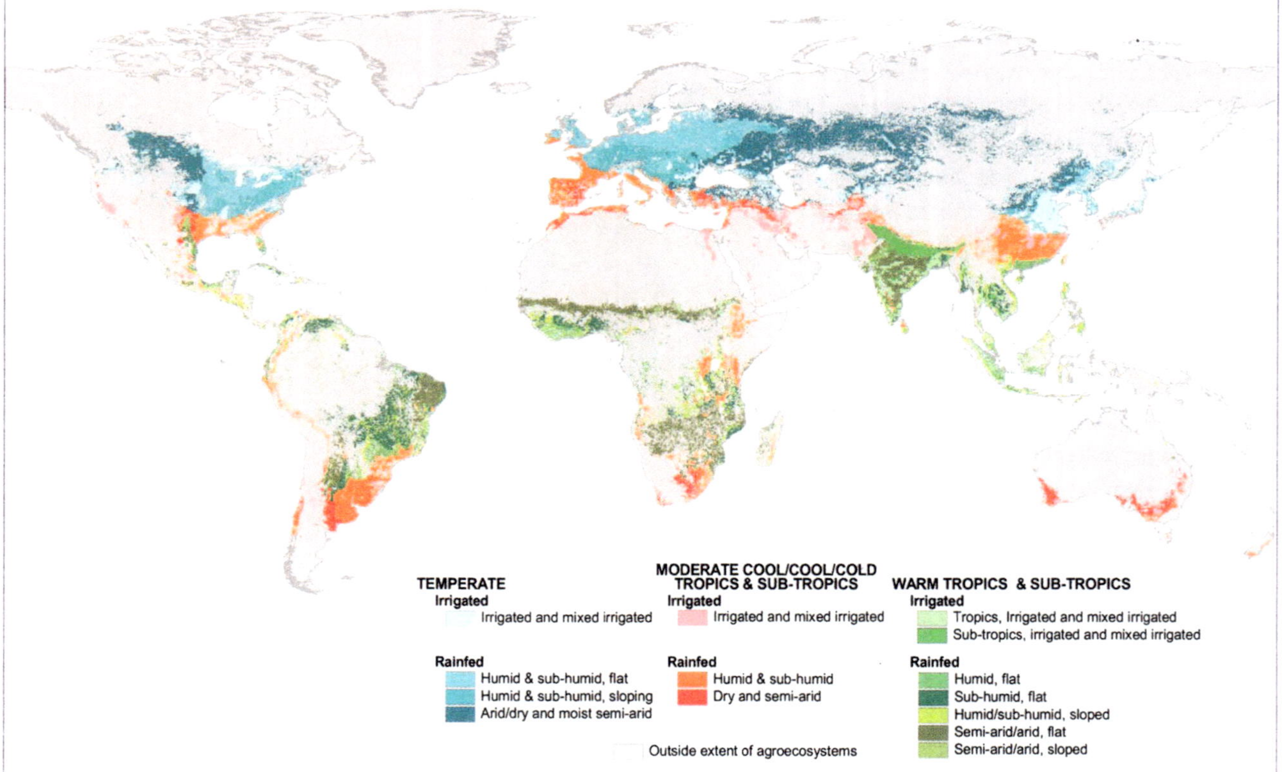

See the section *Agroecologies and Research Spillovers* (overleaf) for a discussion of Figures 6 and 7; see Box 2 (page 16) for a discussion of Figure B2.

Figure B2 Potential drought risk for rainfed production across all cultivated land in agriculture

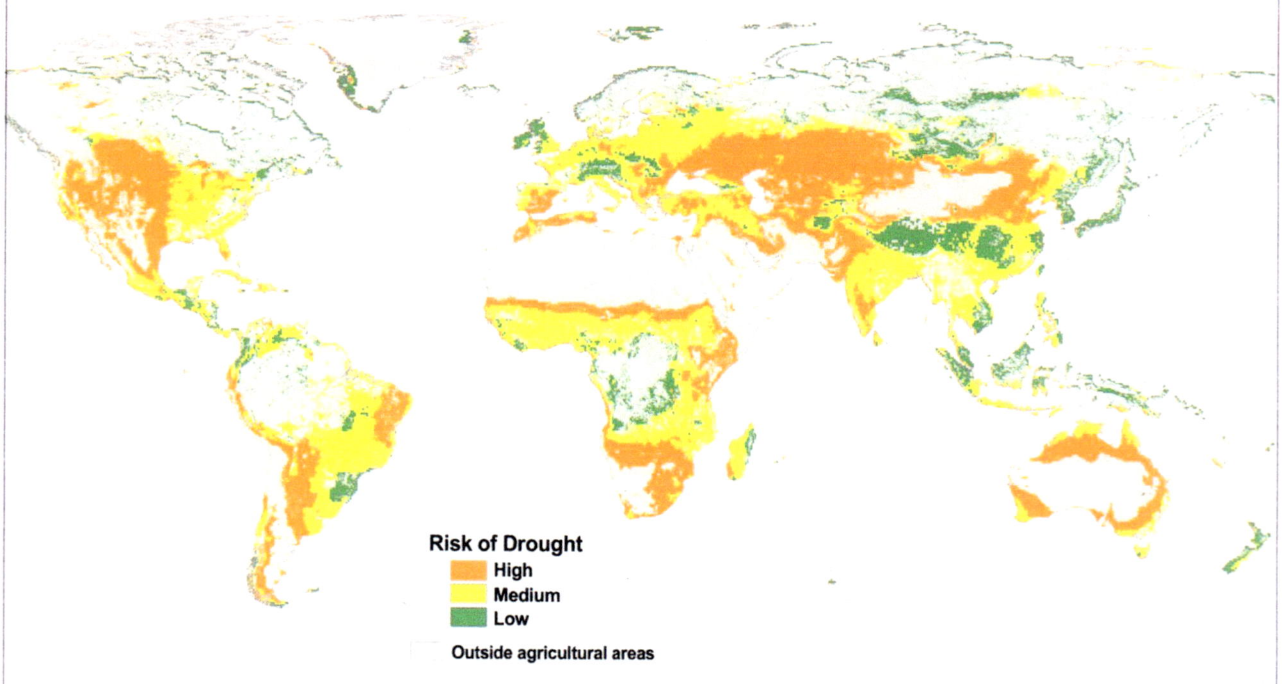

Figure 7 Agroecologies and agricultural production

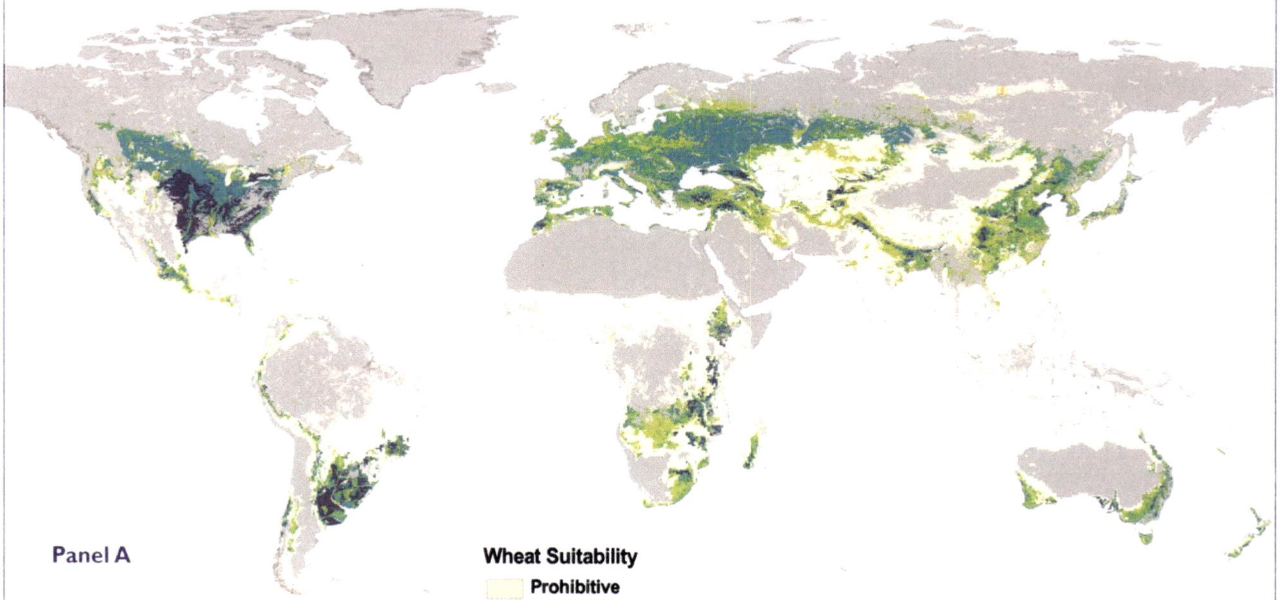

Panel A

Wheat Suitability
- Prohibitive
- Not suitable
- Marginally suitable
- Moderately suitable
- Suitable
- Very suitable

Outside agricultural areas

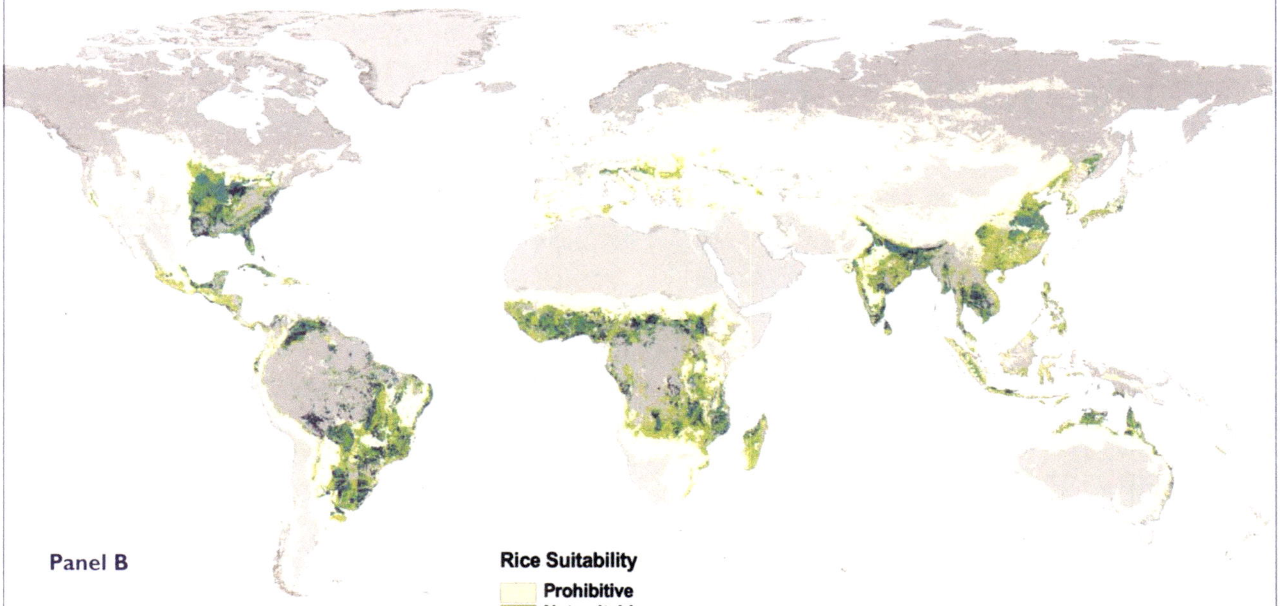

Panel B

Rice Suitability
- Prohibitive
- Not suitable
- Marginally suitable
- Moderately suitable
- Suitable
- Very suitable

Outside agricultural areas

Agroecologies and Research Spillovers

The spending figures previously presented refer to national investments, but agricultural innovation need not be homegrown. A striking feature of the history of agricultural development is that agricultural science and technology spillovers have been pervasive both within and among countries.[18] The result is that agricultural technologies move across borders, both by design and by accident. Spillovers extend beyond agricultural technologies that can be adapted to local conditions to include the underlying knowledge and scientific research.

Most agricultural technologies are sensitive to local climate, soil, and other biophysical attributes, making them less easily transferable than other types of technologies, such as those arising from the medical or information sciences. For example, soybeans are day-length sensitive, so different varieties must be developed for different latitudes. Likewise, many tropical soils are naturally acidic, a less prevalent problem in temperate areas; consequently, crops that thrive in temperate soils can fail or falter under tropical conditions. Variability in the agroecological basis of agriculture means that imported technologies often have to be adapted to local conditions before they can be used (as was usually the case with Green Revolution wheat and rice varieties). Nevertheless, for some developing countries and for some types of technologies, the least-cost option has been to import and adapt technology—and this will continue to be so.

However, while the importance of technology spillover is well recognized, it has often proved difficult to incorporate technology transfer potentials into strategic research-planning perspectives. In part, this simply stems from the limited (informed) use of new sources of data on the distribution of key biophysical attributes of the world's agricultural production environments. Figure 6 (see page 12) provides an agroecological typology of the world's cultivated systems, going beyond the purely rainfall and temperature attributes that, for example, underpinned the agroclimatic characterization by the Food and Agriculture Organization of the United Nations (FAO 1978–81) and the early agroecologically based efforts of the CGIAR (CGIAR/TAC 1991). The typology shown here extends these prior efforts by adding the important distinction between irrigated and rainfed lands, and sloping and flat lands. Furthermore, the map focuses attention on the *actual* rather than *potential* area of cultivated production. Such attributes bring greater geographic specificity to the search for homologous production conditions as a basis for spillins (looking for potential sources of improved knowledge and technology from locations with similar production environments that could be applied locally) or spillouts (taking innovative ideas and technologies known to be successful locally and searching for locations with similar production environments to which the innovations might be transferable).

The regional distribution of agroecological attributes of the world's cultivated systems is presented in the top half of Table 7. Despite the highly aggregated evidence presented, these data suggest scope for potentially significant technology spillover possibilities.[19] For example, the moderately cool tropics and subtropical areas that typify some 14.5 percent of the cultivated area in Sub-Saharan Africa also comprise significant shares of the cultivated systems of Brazil, China, and even the United States. Similarly, the warm tropics and subtropics—flat, rainfed areas that form the greater share of the area in Sub-Saharan Africa—also represent a significant share of the cultivated agriculture in Brazil and India. Increasingly, specific screening criteria—by adding soil characteristics, climate variability, and the like—could be applied in order to delineate increasingly focused geographic domains that might offer opportunities for technology spillover.

There are several ways of exploring the use of spatially explicit information to help guide the search for inward or outward looking technology spillover opportunities. Figure 7 (see page 13) illustrates the spatial incidence of different biophysical suitabilities for the rainfed production of spring wheat (Panel A) and rice (Panel B). Again, being more specific about the attributes of production subsystems and technologies (for example, saline-tolerant lowland rice varieties) would allow the geography of potential spillover opportunities to be more sharply defined.[20] An interesting feature of the two panels, even with this highly aggregated agroecological characterization, is the distinct spatial pattern in the potential production geographies of the two crops. Areas evidently suitable for spring wheat, a

Table 7— Agroecological and production attributes of the world's cultivated systems by region

Region/country	Temperate		Cool tropics and subtropics		Warm tropics and subtropics		
	Rainfed (humid and subhumid) and irrigated	Rainfed (semiarid and arid)	Rainfed (humid and subhumid) and irrigated	Rainfed (semiarid and arid)	Rainfed (humid and subhumid) and irrigated	Rainfed (semiarid and arid)	Total
Share of agricultural area (percent)							
Latin America and the Caribbean	0.0	0.0	26.7	4.2	54.1	14.9	100
Brazil	0.0	0.0	15.0	0.0	68.5	16.5	100
Asia-Pacific	15.0	7.6	21.3	5.2	41.4	9.6	100
China	34.8	19.1	37.8	0.5	7.8	0.0	100
India	0.0	0.0	4.9	0.0	58.8	36.2	100
Middle East and North Africa	6.9	14.4	49.5	29.3	0.0	0.0	100
Sub-Saharan Africa	0.0	0.0	15.9	3.4	37.9	42.8	100
Eastern Europe	33.4	65.4	0.6	0.6	0.0	0.0	100
Developed countries[a]	54.1	13.9	21.8	7.5	1.3	1.4	100
Japan	84.4	2.2	13.5	0.0	0.0	0.0	100
United States	57.3	14.6	18.6	4.1	2.0	3.3	100
World	21.6	16.1	19.2	4.6	26.4	12.2	100
Share of agricultural production (percent)							
Latin America and the Caribbean	0.0	0.6	6.0	7.2	73.4	12.8	100
Brazil	0.0	0.0	1.7	0.0	88.9	9.4	100
Asia-Pacific	26.0	8.4	5.3	1.4	51.4	7.5	100
China	43.0	15.7	8.6	2.2	30.5	0.0	100
India	0.0	0.0	1.1	0.2	73.0	25.7	100
Middle East and North Africa	3.2	12.8	5.5	16.9	36.1	25.4	100
Sub-Saharan Africa	0.0	0.0	16.6	7.5	34.6	41.3	100
Eastern Europe	49.6	48.6	0.2	0.4	0.7	0.4	100
Developed countries[a]	61.7	9.6	2.1	2.7	18.5	5.4	100
Japan	78.4	0.6	0.0	0.0	21.0	0.0	100
United States	61.4	15.1	0.9	4.5	15.0	3.2	100
World	32.6	9.8	4.8	3.6	40.2	9.0	100

SOURCES: Constructed by authors using data and digitized maps underlying Wood, Sebastian and Scherr (2000) for area data and Cassman et al. (2005) for value of production data.

NOTES: The data underlying this table involve a mix of timeframes and spatial scales. The most comprehensive global assessment of cropland extent is from 1992/93, the most complete set of commodity prices in international units is from 1989/91, and the production quantity data are from 1999/2001. The underlying spatial data resolutions range from 1 to 2,500 square kilometers (land cover to climate data, respectively). The global extent of cultivated area is based on a 1 kilometer resolution satellite-derived dataset that contains over 900 million pixels, of which 215 million (23 percent) represent land area and just over 50 million constitute the approximately 37.3 million square kilometers with crop cultivation (areas in which at least 30 percent or more of a pixel is determined to be cultivated).

[a] Includes Organization for Economic Co-Operation and Development (OECD) countries, except Mexico.

crop typically better adapted to temperate climates, do not coincide spatially with those (largely tropical and subtropical) areas suitable for growing rainfed rice.

The broad global extent of cultivated lands depicted in Figure 6 (whose limits are inferred from satellite-derived data on actual land cover, and whose agroecological composition is summarized in the top half of Table 7), represents the outcome of a complex interaction between agroecological (biophysical) factors that condition production *potential* (as indicated for wheat and rice in Figure 7) and a host of socioeconomic, demographic, cultural, and policy factors that have shaped the *realization* of that potential. The bottom half of Table 7 summarizes a novel dataset on the actual spatial incidence of agricultural (that is, crop and livestock) production at the turn of the millennium. Production is summarized in terms of aggregate value, derived as the product of location-specific estimates of production quantities (annual average for 1999–2001) weighted by average world commodity prices (annual average for 1989–91) denominated in international dollars for FAO's production database commodities (Cassman et al. 2005).

Box 2

DROUGHT-INDUCED PRODUCTION RISKS

Droughts manifest themselves in many ways. Of greatest consequence to farmers are agricultural droughts—that is, conditions of water shortage in the root zone of crops during growing season. The crop productivity consequences of drought are most affected by the timing (seasonal distribution) and quantity of rainfall, potential evapotranspiration rates, and the water-holding capacity of soils. Important too are the water use characteristics of individual crops, such as the depth and effectiveness of crop roots and the susceptibility of the crop to water stress at different phenological stages of growth. About 32 percent of the 99 million hectares of wheat grown in developing countries in the early 1990s was exposed to various intensities of drought stress (Rajaram, Braun, and van Ginkel 1996). Thus, crop breeders often test their breeding material for tolerance to water stress at the early, mid, and late stages of the growing season. Different crops, and even different varieties within single crop species, can have quite distinct susceptibilities to drought. All of this makes it difficult to find compact ways of characterizing the production risks associated with different types of agricultural drought. Ideally, measures of drought risk should be developed for specific locations that also reflect the use of specific germplasm in specific production systems. However, the complexity of drought provides broad scope for developing mitigation strategies including selection or development of late planting or early maturing varieties, altered plant architecture to enhance root performance or reduce stomatal release of water, and improved agronomic practices that increase water availability in the rooting zone or minimize heat stress (for example, see Serraj et al. 2003 and Rajaram, Braun, and van Ginkel 1996).

A global perspective of the incidence of regional and global drought risk is summarized in Table B2 and mapped in Figure B2[a] *(for map see page 12)**. The measure of drought risk used is the annual variation (around a three-decade average) of the length-of-growing-period (LGP) computed for each year from 1961 to 1990 (Fischer et al. 2002), arbitrarily divided into three drought-risk classes of roughly equal area globally. This measure is not ideal in terms of representing the multidimensional nature of agricultural drought described above, but it does have certain desirable attributes.[b] Perhaps most importantly, it is derived by taking a long time-series of actual rainfall and evapotranspiration data as input to a soil moisture model that accounts for both the depth and water holding properties of local soils. These calculations were performed worldwide across a 20 by 20 kilometer grid for the climate data and a 5 by 5 kilometer grid for the soils. In each year, the growing season is defined to start when rainfall exceeds half potential evapotranspiration and to conclude when the soil moisture reserve is depleted. From a crop growth perspective, this approach provides a much more relevant measure of drought than could be obtained, for example, by using rainfall records alone. While the measure does not give a clear indication of the type of agricultural drought that might have occurred (for example, early, mid-, or late growing season), variation in the length of growing season incorporates the effects of each, or any combination, of these individual drought types.*

One striking feature of Figure B2 is the geographically widespread extent of potentially drought-prone areas. In some places rainfed production risks can be mitigated by irrigation (separate breakdowns show that drought risks are higher in predominantly irrigated compared with predominantly rainfed areas, and in sloping compared with flat lands). In general, high drought risk is two to four times more common in drier cropland areas, and in the most drought-prone group of agroecosystems—namely, cool/cold, semi-arid, tropics, and subtropics—around three-quarters of the cropland extent is deemed to have highly variable rainfall (although, collectively these agroecosystems represent only 5 percent of global cropland). The second most drought-prone agroecosystem is the drier temperate croplands (comprising some 16 percent of the global cropland total) where almost 70 percent of the area is subject to a high risk of drought. About two-thirds of the cropland of Eastern Europe falls into this category. The least drought-impacted agroecosystem, the warm humid tropics and subtropics, constituting over one-quarter of cropland globally, suffers from highly variable rainfall on only 10 percent of its extent. From a regional perspective, the Middle East/North Africa and Eastern Europe regions appear the most affected, with over half their cropland exposed to high drought risk.

Table B2 reveals some surprising results. Over 90 percent of the dry (arid, semi-arid) temperate cropland in the United States (which is almost 15 percent of U.S. cropland) is categorized as high water-stress risk. Overall, more than 40 percent of U.S. cultivated lands appears to be exposed to inherently high rainfall variability. While reference to the long-term mean LGP must be an integral part of an interregional comparison of water stress conditions, clearly investments in irrigation and well-targeted agronomic practices play a large role in mitigating the potentially negative impacts of this variability.[c] In Sub-Saharan Africa, a region plagued by drought and famine, the results suggest that of the two largest agroecosystems—the humid (38 percent) and the dry (43 percent) warm tropics and subtropics, together representing over 80 percent of the region's total cropland—only 4 percent and 30 percent, respectively, exhibit high water-stress variability. Again, lower average LGP conditions are clearly associated with greater variability in water availability. One potential source of these intuitively low estimates of the incidence of water stress in the region is the weakness of satellite-derived assessments of the cropland extent in Africa. The

Table B2—Incidence of high drought-risk potential for rainfed production by agroecosystem and region

Region/country	Share of global cultivation	Share of agroecosystem exposed to high drought-risk potential (percent): Temperate		Cool (sub)tropics		Warm (sub)tropics		
		Humid	Dry	Humid	Dry	Humid	Dry	All
Developing countries								
Latin America and the Caribbean	17	0		28	96	11	93	31
Brazil	9	0	0	3	0	8	97	22
Asia–Pacific	26	31	68	20	75	11	25	25
China	9	37	71	0	5	0	0	27
India	7	0	0	45		22	25	24
Middle East and North Africa	4	33	38	62	47	0	0	52
Sub-Saharan Africa	15	0	0	36		4	30	20
Eastern Europe	16	35	66			0	0	56
Developed countries[a]	24	19	79	31	85	55	100	36
Japan	<1	0	0	0	0	0	0	0
United States	10	25	94	40	98	57	100	44
World	100	26	68	32	74	10	46	34
Share of global cultivated land		22	16	19	5	26	12	100

SOURCE: Calculated by authors.

NOTES: Table includes only cultivated agroecosystem areas, not rangelands. See Table 7 for area shares of each region by agroecosystem (physical area share of the region exposed to high risk is the product of the area shares in Table 7 and the high risk share in this table). Cells are blank if area shares reported in Table 7 are less than 1 percent of the regional/country total. A zero entry indicates an area share greater than 1 percent, but there are no high drought risks.

[a] Includes Organisation for Economic Co-Operation and Development (OECD) countries, except Mexico, which is included in the Latin America and Caribbean total.

small and often diffuse agricultural land holdings in both wetter and drier areas are poorly discriminated by the type of medium-resolution sensors used to derive the global land cover data used in this analysis. However, as shown in Figure B2, a broader perspective of agricultural land, including pasture and grazing areas (which undoubtedly contain many cropped holdings undetected by the satellite sensors) together with areas designated as cropland, reveals that a large share of the agricultural area of the Sahel and eastern and southern Africa is clearly predisposed to high drought risk.

SOURCES: Fischer et al. 2002; Hayes et al. 2004; Rajaram, Braun, and van Ginkel 1996; Serraj et al. 2003.

[a] *The map indicates growing period variability over all agricultural land (cropland, rangeland, and pasture), whereas the table summarizes this variability within cropland only (as depicted in Figure 6).*

[b] *The key shortcoming is that the variability (coefficient of variation, or CV) in LGP encompasses both longer (wetter) as well as shorter (drier) growing seasons in the historic record. Furthermore, variability presents less of a production risk where LGP is high. Thus a +/– 20 percent variability in LGP likely presents less production risk (certainly less risk of total harvest failure) where annual average LGP is 270 days than where it is 90 days. Empirically, however, visual inspection of Figure B2 suggests that the CV of LGP does provide an intuitively reasonable picture of areas that are more or less drought prone.*

[c] *Though precise economic impacts are difficult to assess, the U.S. National Drought Mitigation Center at the University of Nebraska has compiled a range of estimates, including some US$11 billion dollars in losses to agriculture as a consequence of the 2002 drought. The center also reports indemnities paid for crop losses (predominantly for drought) in Nebraska alone were $190 million for the drought of 2000, and more than $372 million for the drought of 2002 (Hayes et al. 2004).*

Significant differences are found in the spatial patterns of agricultural areas and the value of agricultural production. Whereas more than 35 percent of the world's agricultural area is found in the semi-arid/arid temperate zones and the humid/subhumid cool tropics/subtropics (columns 2 and 3 of Table 7, respectively) these zones contribute only 15 percent of the world's agricultural production value. It is estimated that over 40 percent of the value of agricultural output comes from the humid/subhumid warm tropics/subtropics (column 5 in Table 7, a zone that accounts for only 26 percent of the world's agricultural area), with a further 33 percent of agricultural value generated in the humid/subhumid temperate zone (column 1 in Table 7). Notably, both these tropical and temperate zones include substantial irrigated areas. This discordance between the physical area allocated to agriculture and the value of production each area generates persists across most regions of the world. An important implication of these new data is that it could be misleading to use area allocation as the sole production-related criterion for targeting agricultural research, thereby ignoring the agroecological and economic realities that clearly have a dominant effect on the spatial distribution of production value.

Importantly, both the supply and demand for spillover technologies appears to be changing. Notably, rich countries are reorienting their agricultural R&D away from the types of technologies that are most easily adapted and adopted by developing countries (Pardey, Alston, and Piggott 2006). In addition, intellectual property rights and other regulatory policies—including biosafety protocols, trading regimes, and specific regulatory restrictions on the movement of genetic material—are increasingly influencing the extent to which such spillovers are feasible or economic.

Some developing countries have expanded their own research capacity and shifted upstream, reducing their emphasis on adaptive R&D (examples include the largest developing countries: Brazil, China, and India). These countries have become a potential source of new technologies for the poorest and smallest countries, which will (or often should, given economic realities in the current and foreseeable future) continue to emphasize adaptive research.

CGIAR Trends

While the CG system has captured the attention of the international agricultural R&D and aid communities through the impact of its scientific achievements and its pivotal role in the Green Revolution, it has spent only a small fraction of the global agricultural R&D investment. In 2000, the CG represented 1.5 percent of the $23 billion (2000 prices) global public-sector investment in agricultural R&D and just 0.9 percent of all public and private agricultural R&D spending.

The nominal and real (that is, adjusted for inflation) values of total expenditures for the CGIAR are shown in Figure 8. The CG system began modestly. Between 1960 and 1964, of the institutes that would become the CG, only IRRI was operating as such. After an initial expenditure of US$7.4 million in 1960, total spending rose to $1.3 million per year in 1965. By 1970, the four founding centers—IRRI, CIMMYT, IITA, and CIAT—were allocated a total of $14.8 million annually.[21] The progressive expansion of the total number of centers and the funding per center during the next decade involved a 10-fold increase in nominal spending, to $141 million in 1980. During the 1980s, spending continued to grow, more than doubling in nominal terms to reach $305 million in 1990. The rate of growth had slowed but was still impressive. In the 1990s, however, although the number of centers grew—from 13 to 18 at one point, but now 15—funding did not grow enough to maintain the level of spending per center, let alone the growth rates. Since 2000, funding has grown in total but with a continuing trend toward earmarked support for specific projects and programs of research involving multiple centers and other research providers outside the CG.

In the early years of plenty, all the centers grew together, but even during the bountiful 1970s and 1980s, when all of the centers grew, they did not all grow at the same rate. A notable trend has been the declining share of the four founding centers. In 1971, these four centers accounted for 100 percent of the allocation. By 1980, their share had slipped to 54 percent, and by 2004 it was down to 36 percent. During the stagnation of the 1990s, nine centers experienced a nominal decline in support, including the four founders—IRRI, CIAT, CIMMYT, and IITA—along with CIP, ICRISAT, ILRI (formerly ILCA and ILRAD), ICARDA, and ISNAR.[22] The centers being downgraded also tended to be the larger centers. Among the pre-1990 centers, the International Plant Genetic Resources Institute (IPGRI) grew the fastest, with its funding more than doubling in just five years. Of the new entrants, the two forestry institutes showed the greatest gains. These broad trends indicate that, through both the addition of new centers and the allocation of funds among centers, the agenda of the CG shifted dramatically away from its original focus, especially in the 1990s.

In its early years, virtually all CG funding came in the form of unrestricted support (wherein the funds were earmarked by center and spending within a center was largely at the discretion of center management). This remained the dominant mode of funding for the CGIAR

Figure 8 Nominal and real expenditures of CGIAR-supported centers

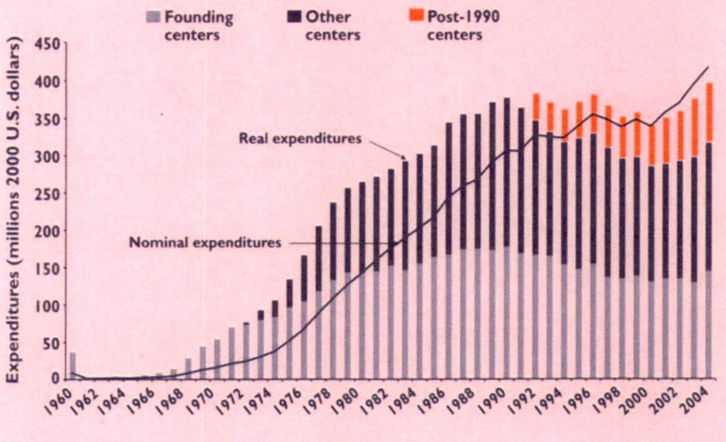

SOURCE: Alston, Dehmer, and Pardey (2006).

NOTES: Expenditures are in 2000 prices. Pre-1972 expenditures represent funds to precursor international research institutes. Data for the International Rice Research Institute (IRRI) date from 1960; data for the International Center for Maize and Wheat Improvement (CIMMYT) and Centro Internacional de Agricultura Tropical (CIAT) date from 1966; and data for International Institute of Tropical Agriculture (IITA) date from 1971.

throughout the 1970s, a period when total funding grew the fastest and the number of centers grew from 4 in 1971 to 13 in 1980 (Figure 8). Typically, new entrants were fully funded with unrestricted support, while unrestricted funding for existing centers remained a very significant share of their respective totals throughout the 1970s and early 1980s. For example, in 1982, unrestricted funds as a share of the CG total averaged 84 percent—about 82 percent for the four founding centers and 87 percent for the incoming centers.

The period after 1983 was one of a continuing decline in the share of unrestricted funds—down to 44.5 percent of the total in 2003 compared with a 1980s average of 81.2 percent (and a 1970s average of 88.2 percent for the precursor centers of the CG). This decline since the early 1980s has two distinct phases. For the period through to 1987, the unrestricted share fell, while total funding for the CG (in real terms) continued to rise. For the period thereafter, both real funding and the unrestricted share declined, partly reflecting the fact that most of the newly admitted centers in the 1990s joined with comparatively small shares of unrestricted support (unlike the wave of new entrants that joined during the 1970s). The corollary to the decline in the share of unrestricted funding is a rise in the share of funds earmarked for specific purposes.

Development Aid and Agricultural R&D

Development aid as a source of funding agricultural R&D has been an important source of sponsorship for CGIAR research and has also played a pivotal part in underwriting national R&D efforts in some countries and some regions of the world, especially in Sub-Saharan Africa. Piecing together a coherent picture of general trends in the aid–agricultural R&D relationship is hampered by access to data; hence relevant, yet incomplete, evidence is presented below.

Since 1960, official development assistance (ODA) from the Development Assistance Committee (DAC) countries, including both multilateral and bilateral assistance, grew in real terms to a peak of US$72.6 billion (2000 prices) in 1992, dropping to $51.2 billion by 2001 (and increasing thereafter to $74.5 billion in 2004). There was no clear shift in the bilateral ODA share over time—during the 1990s, bilateral (country-to-country) aid averaged around 70 percent of total aid (Table 8). Data on the sectoral orientation of aid are available for bilateral but not multilateral funds. In contemporary times, the agricultural component of bilateral assistance grew steadily, to peak at $6.5 billion in 1988 and decline thereafter to just $2.0 billion in 2003. The data suggest a strong shift away from agriculture in aid funding priorities. As a share of all bilateral aid, agriculture fell from 15.2 percent in 1988 to only 4.2 percent in 2003.

Aid for Agricultural R&D

After several decades of strong support, international funding for agriculture and agricultural research began to decline around the mid-1980s, as support for economic infrastructure as well as health, education, and other social services began to grow. Data on aggregate trends are simply not available, but information on agricultural R&D grants and loans from the World Bank and the United States Agency for International Development (USAID) is obtainable.

The amount of funding that USAID directed toward agricultural research conducted by national agencies in less-developed countries declined by 75 percent in inflation-adjusted terms from the mid-1980s to 2004 (Figure 9, Panel A). Asian countries experienced the largest losses, from around $45 million (in 2000 prices) in the mid-1980s to zero in 2004. Support to Africa and

Table 8—Aid to agriculture, 1970–2004

Year	Total official development assistance (ODA) (million 2000 U.S. dollars)	Bilateral aid: Amount (million 2000 U.S. dollars)	Bilateral aid: Share to agriculture (percent)
1970	24,719	20,886	4.91
1975	35,448	26,233	11.13
1980	49,166	31,875	16.63
1985	41,773	30,782	15.93
1990	67,071	47,540	11.39
1995	64,077	44,129	9.82
2000	53,749	36,064	6.36
2003	65,502	47,222	4.22
2004	74,483[a]	50,700[a]	n.a.

SOURCES: Adapted from Alston, Dehmer, and Pardey (2006).

NOTE: n.a. indicates not available.

[a] Preliminary estimate.

Latin America and the Caribbean was also cut severely: by 2004 funding had fallen to only 23 percent of mid-1980s levels for Africa and 2 percent for Latin America and the Caribbean. Since 1997, USAID funding for *all* of agriculture (including agricultural R&D) has also failed to regain the ground it lost. Thus funding for agricultural R&D has suffered, with the majority going to support global (mainly CGIAR) research rather than research conducted in national labs in the developing world. In fact, in 2004, the data underpinning Figure 9a indicate that USAID committed just $15 million to nationally performed agricultural R&D for the *whole* of Sub-Saharan Africa.

Over the past two decades, World Bank lending to the rural sector has been erratic, but after adjusting for inflation, the general trend has been downward. Agriculture's share of total lending has also declined (from an average of 26 percent during the first half of the 1980s to only 10 percent by 2000). There is no discernable pattern in the amount of World Bank lending authorized for agricultural R&D, other than a temporary increase in loan approvals in the late 1980s and early 1990s, and an exceptionally large amount of lending in 1998 resulting mostly from loans with large research components approved for India ($136 million, current prices), China ($68 million), and Ethiopia ($60 million) (Figure 9, Panel B). The size of the loans has been highly variable, ranging from $0.1 million for Argentina in 1992 and Niger in 1997, to $136 million for India in 1998.

Figure 9 Donor support for agricultural R&D by region, in U.S. dollars

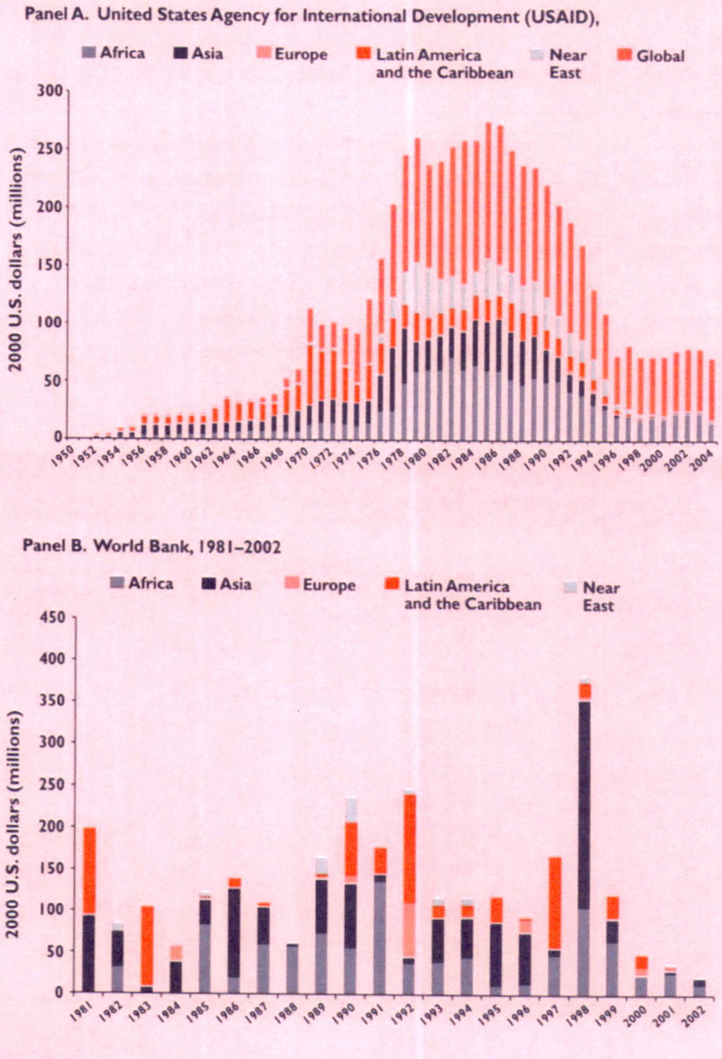

SOURCE: Calculated by authors based on World Bank (2005a) and Alex (2004).

Sub-Saharan Africa

The era of substantial donor support for agricultural R&D in Sub-Saharan Africa appears to be drawing to a close—with certainty if recent trends continue. Donor contributions (including World Bank loans) accounted for an average of 35 percent of funding to principal agricultural research agencies in 2000. Pardey and Beintema (2001) estimated that five years earlier, close to half the agricultural research funding in the region was derived from donor contributions (Figure 10).

These regional averages mask great variation among countries. In 2000, donor funding accounted for more than half of the agricultural R&D funding in 7 of the 23 sample countries. Eritrea, in particular, was highly dependent on donor contributions. Its principal agricultural research agency received more than three-quarters of its funding from donors. In contrast, donor funding was quite insignificant in Botswana, Malawi, Mauritius, and Sudan (less than 5 percent). From the mid-1990s to 2000, one-third of the sample countries experienced declines of 10 percent or more in the donors' share of total agricultural R&D funding, while the share of funding from donor sources increased by at least 10 percent for four countries (namely, Burundi, Gambia, Tanzania, and Togo). Notably, donor funding fell from over 50 percent of the total to 10 percent or less for Malawi, Niger, and Sudan, as a result of the completion of major projects funded by World Bank loans or contributions from the FAO.

Funding from nongovernment or donor sources, such as internally generated revenues, was comparatively small, representing just 11 percent of total funding in 2000. The exceptions are Benin and Côte d'Ivoire. The principal agricultural research agencies in these two countries generated significant shares of total funding from research contracts, commercialization of agricultural products, and dissemination of research results. In some cases, this practice was dictated by the terms of the international loans for agricultural R&D. For example, in Côte d'Ivoire, the World Bank's second National Agricultural Services Support Project (PNASA II) had an important commercialization component, stipulating that 35 percent of the annual budget of the National Agricultural Research Center (CNRA) was to be self-generated through mechanisms such as commodity sales (Beintema and Stads 2004).

Implications

There are substantial and potentially profound changes under way regarding agricultural R&D worldwide. Global investments in agricultural research have continued to grow, albeit at a much slower rate in the 1990s than in previous decades. These new data suggest a global bifurcation in the conduct of agricultural R&D, with a select few developing countries showing signs of closing in on the higher amounts and higher intensity of investment in agricultural R&D typically found in the rich countries. Meanwhile, a large number of developing countries are either stalling or slipping in terms of the amount spent on agricultural R&D, the intensity of investment, or both. The private sector maintains a big presence in the rich countries and a growing presence in a small number of developing countries where agricultural input markets are growing and becoming more cost-effective to service.

In 2000, 80 developing countries in the world got by with a combined total of just $1.4 billion of public agricultural R&D spending (about 6.3 percent of the global total). By way of comparison, more than 35 public universities in the United States each spent in excess of this amount in 2004. Increasing the amount spent on agricultural R&D in low-income countries that are heavily reliant on agriculture is likely to be a wise, but difficult, investment given the pressing demands on the cash-strapped governments in these economies (Runge et al. 2003). However, simply maintaining current agricultural R&D policies could leave many developing countries as agricultural technology orphans in the decades ahead. Developing countries may have to become more self-reliant and perhaps more dependent on one another for the collective goods of agricultural R&D and technology (Pardey, Alston, and Piggott 2006). Some of the more advanced developing countries like Brazil, China, India, and South Korea seem to be gaining ground, with productive and self-sustaining local research sectors taking hold. However, other parts of the developing world, like Bangladesh, Indonesia, and Zambia, are merely regaining lost ground or slipping further behind. Aside from a handful of larger countries, significant numbers of developing countries, especially in Africa, continue to face serious funding and institutional constraints that inhibit the effectiveness of local R&D. Together these factors may spell serious food deficits for some of these countries.

Achieving the rate of agricultural productivity gains necessary to feed the generally faster-than-average growing populations in the poorer parts of the world requires giving much more explicit attention to tapping and adapting technologies developed elsewhere and better targeting of those technologies to maximize local

Figure 10 Sources of funding by country in Sub-Saharan Africa, 1995/96 and 2000

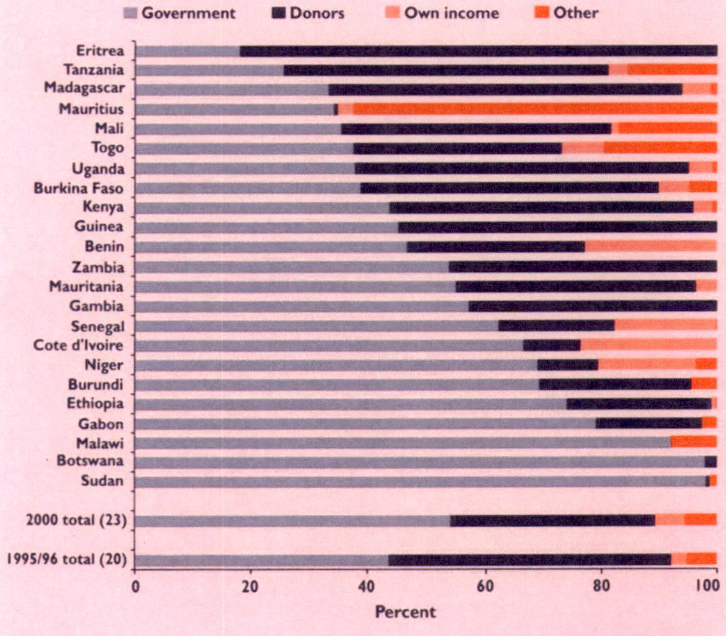

SOURCE: Beintema and Stads (2004).

NOTES: Funding data include only the main agricultural research agencies in each of the respective countries. Combined, these agencies accounted for 76 percent of total spending for the 23-country sample in 2000. Data for West Africa, with the exception of Nigeria, are for 2001.

food-security and agricultural development impact. However, the shifting scientific orientation of rich-country research, combined with changing biosafety and intellectual property regimes internationally, suggests that the technology spillover pathways of the past may not carry forward, even to the near future. New institutional arrangements (including improving the allocation and efficiency with which scarce agricultural R&D funds are deployed) will likely be required—and are possible, if the pipeline of agricultural technologies useful for poor-country farmers is to be kept fully primed.

Notes

1. See Alston et al. (2000) for a comprehensive review.
2. This section draws from Pardey, Dehmer, and El Feki (2006). In measuring and classifying (agricultural) research expenditures, the guidelines laid down in OECD (2002) have been followed. Research spending data are reported by performer, irrespective of the funding source.
3. All currency values are expressed in international dollars unless explicitly stated otherwise. Data in this section refer to gross expenditures on research and development (GERD). See Box 1 for more details on the methods used to deflate R&D spending denominated in current local currency units to 2000 prices and then convert the results to a common numeraire currency unit.
4. Here, and throughout this report, growth rates are calculated by the least squares regression method detailed by the World Bank (2006, 324). IAC (2004) provides a comprehensive assessment of the need for strengthening science capacity throughout the developing world.
5. Notably, during the second half of the 1990s, GERD grew by just 2.2 percent per year in Japan—roughly one-third the corresponding U.S. annual rate of 6.0 percent.
6. Hereafter, the terms developed, rich, and high-income are used interchangeably, referring to the high-income countries of the OECD. Further, the terms developing, poor, and low-income are also used interchangeably, referring to all countries other than the high-income members of the OECD.
7. In 2000, South Africa accounted for nearly 25 percent of all agricultural R&D spending in Sub-Saharan Africa, although Liebenberg and Kirsten (2006) note the severe contraction in research capacity of the Agricultural Research Council (ARC), the largest provider of agricultural research in South Africa (accounting for nearly 58 percent of the country's agricultural R&D spending in 1999). They note that "the number of research staff at ARC dropped from 751 in 1992 to 682 in 2000 (and 525 in April 2003)" Liebenberg and Kirsten (2006, 212).
8. For more details, see Pal and Byerlee (2006), on India, and Fan, Qian, and Zhang (2006), on China.
9. For some purposes it may be useful to classify nonprofit agencies as private entities, distinguishing between private-for-profit and private, nonprofit research.
10. More strictly speaking, this has been the case in western and central Africa, eastern and central Africa, and eastern and southern Africa, with country overlap among these regional groupings. For example, the Democratic Republic of Congo (central Africa) is a member of both the western and eastern subregions, and Tanzania is a member of both eastern and southern consortia.
11. It is likely that the national and international R&D trends presented above (based on a research performers) and below (based on development assistance for research) capture the vast majority of subregional expenditures on agricultural R&D, irrespective of the mode of allocating and conducting the research. The overall trend, at least through to 2000, indicates a continuing slowdown in total support for African agricultural R&D.
12. Cross-country collective action to conceive and conduct research, absent counterpart institutional innovation to generate and allocate funding for regional research, may be of little positive consequence. The Regional Fund for Agricultural Technology (FONTAGRO)—established in 1998 as a competitive, regional mechanism to support research and innovation throughout Latin America and the Caribbean—is a creative example of one such funding mechanism. FONTAGRO has channeled a total of US$9.5 million (not counting $2 million available for the current, 2006, call for proposals) as direct support to projects, with regional organizations and others providing approximately $21 million in counterpart funding. This is well below the funding expectations for this initiative, underscoring the reluctance of national governments (and others) to cede funds for regional research, even if they are the potential beneficiaries of that research. Binenbaum and Pardey (2005), for example, describe a range of complex incentive problems that befall collective action in crop breeding research.

13. This may indicate that increasingly knowledge-intensive agricultural production systems must invest increasingly greater shares of their research in simply maintaining (distinct from increasing upon) past yield gains. See Alston, Norton, and Pardey (1998, 31–32) and the citations therein for more discussion of this point.

14. There are only 26 countries in the Sub-Saharan African sample for this calculation because Eritrea did not exist as such until 1992.

15. In 2000, average spending per capita in Latin America and the Caribbean was also lower than in 1981 but higher than the corresponding 1991 ratio.

16. Alston and Pardey (2006) provide historical data and a more complete description of long-run technological and productivity developments in U.S. agriculture.

17. Wright et al. (2006) provide a more comprehensive treatment of the forms of intellectual property protection used in agriculture worldwide.

18. Assessing the formal impact evaluation literature, Alston (2002) contends that up to half the local productivity gains in agriculture over the past several decades are attributable to the effects of spillin technologies developed elsewhere. For example, Pardey et al. (1996) showed that research conducted on wheat and rice by the International Maize and Wheat Improvement Center (CIMMYT) and the International Rice Research Institute (IRRI), respectively, almost entirely in developing countries, provided very large economic benefits to the U.S. wheat and rice sectors. This was as a consequence of technology spillover, where wheat and rice varieties generated for developing-country farmers could either be adopted directly by U.S. farmers or, more often, be incorporated into U.S.–focused crop improvement programs. See also Evenson and Gollin (2003).

19. The table presents agroecological attributes aggregated to 6 classes. These are derived from the 14 classes shown in Figure 6, which in turn is a summary of more disaggregated attributes. Using geographical information system (GIS) tools, the degree of aggregation of the agroecological data can be varied so as to best match the nature of the technology investment and development decisions being assessed.

20. See Box 2 for a more in-depth assessment of the spatially variable extent and nature of drought in an agricultural context.

21. IRRI is the International Rice Research Institute in the Philippines, CIMMYT the Spanish acronym for the International Maize and Wheat Improvement Center in Mexico, IITA the International Institute of Tropical Agriculture in Nigeria, and CIAT the Spanish acronym for the International Center for Tropical Agriculture in Colombia.

22. CIP is the International Potato Center, ICRISAT is the International Crops Research Institute for the Semi-Arid Tropics, ILRI is the International Livestock Research Institute, ICARDA is International Center for Agricultural Research in the Dry Areas in, and ISNAR is the International Service for National Agricultural Research, now a division of the International Food Policy Research Institute, based in Addis Ababa, Ethiopia.

References

Alex, G. 2004. Estimated USAID funding for agricultural research. Farmer-to-Farmer Program Advisor, United States Agency for International Development, Economic Growth, Agriculture, and Trade Bureau, Washington, D.C. Unpublished data file.

Alston, J. M. 2002. Spillovers. *Australian Journal of Agricultural and Resource Economics* 48 (3): 315–346.

Alston, J. M., and P. G. Pardey. 2006. Agricultural productivity. Subchapter in *Historical statistics of the United States: Earliest times to the present?* Millennial edition, Volume 4, Part D, *Economic sectors*, S. B. Carter, S. S. Gartner, M. R. Haines, A. L. Olmstead, R. Sutch, and G. Wright, eds. Cambridge, U.K.: Cambridge University Press.

Alston, J. M., S. Dehmer, and P. G. Pardey. 2006. International initiatives in Agricultural R&D: The changing fortunes of the CGIAR. Chapter 12 in *Agricultural R&D in the developing world: Too little, too late?* P. G. Pardey, J. M. Alston, and R. R. Piggott, eds. Washington, D.C.: International Food Policy Research Institute.

Alston, J. M., G. W. Norton, and P. G. Pardey. 1998. *Science under scarcity: Principles and practice for agricultural research evaluation and priority setting.* Wallingford, U.K.: CAB International.

Alston, J. M., C. Chan-Kang, M. C. Marra, P. G. Pardey, and T. J. Wyatt. 2000. *A meta-analysis of the rates of return to agricultural R&D: Ex pede Herculem.* IFPRI Research Report No. 113. Washington, D.C.: International Food Policy Research Institute.

Beintema, N. M., and G-J. Stads. 2004. *Investing in Sub-Saharan African agricultural research: Recent trends.* 2020 Africa Conference Brief No. 8. Washington, D.C.: International Food Policy Research Institute.

Binenbaum, E., and P. G. Pardey. 2005. Collective action in plant breeding. Selected paper presented at the American Agricultural Economics Association Annual Meeting, Providence, Rhode Island, July 24–27.

Casas, J., M. Solh, and H. Hafez, eds. 1999. *The national agricultural research systems in the West Asia and North Africa region.* Aleppo, Syria: International Center for Agricultural Research in the Dry Areas (ICARDA), Food and Agriculture Organization of the United Nations, Association of Agricultural Research Institutions in the Near East and North Africa (AARINENA), and Mediterranean Agricultural Institute (CIHEAM).

Cassman, K. G., S. Wood, Poh Sze Choo, H. D. Cooper, C. Devendra, J. Dixon, J. Gaskell, S. Khan, R. Lal, L. Lipper, J. Pretty, J. Primavera, N. Ramankutty, E. Viglizzo, K. Wiebe, S. Kadungure, N. Kanbar, Z. Khan, R. Leakey, S. Porter, K. Sebastian, and R. Tharme. 2005. Cultivated systems. Chapter 26 in *Ecosystems and human well-being: Current states and trends*, R. Hassan, R. Scholes, and N. Ash, eds. Washington, D.C.: Island Press.

CGIAR/TAC (Technical Advisory Committee of the Consultative Group on International Agricultural Research). 1991. *An ecoregional approach to research in the CGIAR.* Rome: TAC Secretariat.

Eurostat. 2005. Science and technology. <http://epp.eurostat.cec.eu.int/portal/page?_pageid=0,1136250,0_45572558&_dad=portal&_schema=PORTAL> (accessed May 2005).

Evenson, R. E., and D. Gollin. 2003. *Crop variety improvement and its effect on productivity.* Wallingford, U.K.: CAB International.

Fan, S., K. Qian, and X. Zhang. 2006. China: An unfinished reform agenda. Chapter 3 in *Agricultural R&D in the developing world: Too little, too late?* P. G. Pardey, J. M. Alston, and R. R. Piggott, eds. Washington, D.C.: International Food Policy Research Institute.

FAO (Food and Agriculture Organization of the United Nations). 1978–81. *Report of the agroecological zones project.* World Soils Resources Report No 4. Volumes 1–4. Rome.

_________. 2005a. FAOSTAT database. <http://faostat.fao.org/> (accessed September 2005).

_________. 2005b. *FAO statistical yearbook.* Volume 1. Rome.

Fischer, G., H. van Velthuizen, M. Shah, and F. Nachtergaele. 2002. *Global agro-ecological assessment for agriculture in the 21st century: Methodology and results.* RR-02-02. Laxenburg, Austria: International Institute for Applied Systems Analysis.

Hayes, M. J., M. D. Svoboda, C. L. Knutson, and D. A. Wilhite. 2004. Estimating the economic impacts of drought. Selected paper at the joint 14th Conference on Applied Climatology/15th Symposium on Global Change and Climate Variations, Seattle, January 11–15, 2004.

IAC (Inter-Academy Council). 2004. *Inventing a better future: A strategy for building worldwide capacities for science and technology.* Amsterdam: Royal Netherlands Academy of Arts and Sciences.

Liebenberg, F., and J. Kirsten. 2006. South Africa: Coping with structural changes. Chapter 8 in *Agricultural R&D in the developing world: Too little, too late?* P.G. Pardey, J. M. Alston, and R. R. Piggott, eds. Washington, D.C.: International Food Policy Research Institute.

NSF (National Science Foundation). 2005. *National patterns of research and development resources: 2003*, NSF 05-308. Arlington, VA, U.S.A.: National Science Foundation, Division of Science Resources Statistics.

OECD (Organisation for Economic Co-Operation and Development). 2002. *Frascati manual: Proposed standard practice for surveys on research and experimental development.* Paris.

_________. 2005. *OECD science and technology indicators 2005.* Paris.

Pal, S., and D. Byerlee. 2006. India: Evolution and emerging policy issues. Chapter 7 in *Agricultural R&D in the developing world: Too little, too late?* P. G. Pardey, J. M. Alston, and R. R. Piggott, eds. Washington, D.C.: International Food Policy Research Institute.

Pardey, P. G., and N. M. Beintema. 2001. *Slow magic: Agricultural R&D a century after Mendel.* IFPRI Food Policy Report. Washington, D.C.: International Food Policy Research Institute.

Pardey, P. G., J. M. Alston, and R. R. Piggott, eds. 2006. *Agricultural R&D in the developing world: Too little, too late?* Washington, D.C.: International Food Policy Research Institute.

Pardey, P. G., S. Dehmer, and S. El Feki. 2006. *Global spending on science: A new order in the making?* International Science and Technology Practice and Policy (InSTePP) Center. St Paul: University of Minnesota (in preparation).

Pardey, P. G., J. Roseboom, and B .J. Craig. 1992. A yardstick for international comparisons: An application to national agricultural research expenditures. *Economic Development and Cultural Change* 40 (2): 333–349.

Pardey P. G., J. M. Alston, J. E. Christian, S. Fan. 1996. *Hidden harvest: U.S. benefits from international research aid.* IFPRI Food Policy Report. Washington, D.C.: International Food Policy Research Institute.

Rajaram, S. R., H. J. R. Braun, and M. R. van Ginkel. 1996. CIMMYT's approach to breed for drought tolerance. *Euphytica* (historical archive). 92 (1–2): 147–153.

RICYT (Red de Indicadores de Ciencia y Technologîa). 2005. Network on science and technology indicators. <http://www.ricyt.org/default.asp?Idioma=ENG> (accessed May 2005).

Runge, C. F., B. Senauer, P. G. Pardey, and M. W. Rosegrant. 2003. *Ending hunger in our lifetime: Food security and globalization.* Baltimore: Johns Hopkins University Press.

Serraj, R., F. R. Biddinger, Y. S. Chauhan, N. Seetharama, S. N. Nigam, and N. P. Saxena. 2003. Management of drought in ICRISAT cereal and legume mandate crops. Chapter 8 in *Water productivity in agriculture: Limits and opportunities for improvement*, J. W. Kine, R. Barker and D. Molden, eds. Wallingford, U.K.: CAB International.

USDA/CRIS (United States Department of Agriculture, Current Research Information System). 2006. Unpublished data. USDA, Washington, D.C.

Wood, S., K. Sebastian, and S. J. Scherr. 2000. *Pilot analysis of global ecosystems: Agroecosystems.* Washington, D.C.: International Food Policy Research Institute and World Resources Institute.

Wright, B. D., P. G. Pardey, C. Nottenburg, and B. Koo. 2006. Agricultural innovation: Economic incentives and institutions. In *Handbook of agricultural economics*, Volume 3, R. E. Evenson, P. Pingali, and T. P. Schultz, eds. Amsterdam: Elsevier (forthcoming).

World Bank. 2005a. World Bank agricultural research by region. Washington, D.C. Unpublished data file.

________. 2005b. World development indicators. <http://www.worldbank.org/data/onlinedatabases/onlinedatabases.html> (accessed May 2005).

________. 2006. *The world development report 2006: Equity and development.* New York: Oxford University Press and the World Bank.

Made in the USA
Monee, IL
05 November 2026

67379783R00024